T0353509

Understanding our Unseen Reality

Solving Quantum Riddles

Understanding our Unseen Reality

Solving Quantum Riddles

Ruth E. Kastner

University of Maryland, USA

ICP

Imperial College Press

Published by

Imperial College Press
57 Shelton Street
Covent Garden
London WC2H 9HE

Distributed by

World Scientific Publishing Co. Pte. Ltd.

5 Toh Tuck Link, Singapore 596224

USA office: 27 Warren Street, Suite 401-402, Hackensack, NJ 07601

UK office: 57 Shelton Street, Covent Garden, London WC2H 9HE

Library of Congress Cataloging-in-Publication Data
Kastner, Ruth E., 1955– author.
Understanding our unseen reality : solving quantum riddles / Ruth E. Kastner, University of Maryland, USA.
pages cm
Includes bibliographical references and index.
ISBN 978-1-78326-695-1 (hardcover : alk. paper) -- ISBN 978-1-78326-646-3 (pbk. : alk. paper)
1. Quantum theory. I. Title.
QC174.125.K37 2015
530.12--dc23
2014044002

British Library Cataloguing-in-Publication Data
A catalogue record for this book is available from the British Library.

Typeset by Stallion Press
Email: enquiries@stallionpress.com

Printed in Singapore

Preface

If you think that there is more to the universe than what we can see, and that we can gain a better understanding of that unseen world even if we cannot directly observe it with our five senses, then this book was written for you. We are on the verge of a scientific revolution, and this book is an effort to bring these revolutionary ideas to the interested reader.

The specific aim of this book is to present an accessible account of my extended version of the Transactional Interpretation of Quantum Mechanics (TIQM), first proposed by Prof. John G. Cramer. No background in mathematics or physics is assumed; the only requirement is a healthy curiosity and, as noted above, an open mind. While many popular books on quantum theory do a good job of laying out the perplexities and unsolved riddles of quantum theory, this book offers some specific solutions to those riddles. The solutions involve not only the transactional picture, but also a paradigm change: we can no longer think of reality as confined to the arena of space and time. Reality extends beyond the observable realm of space and time, and quantum theory is what describes those extended but hidden aspects.

The transactional interpretation (TI) was discussed previously in the popular science genre by John Gribbin in his book *Schrödinger's Kittens and the Search for Reality* (1995). Shortly after the publication of Gribbin's book, philosopher Tim Maudlin (2002) raised an objection to TIQM in the philosophical literature which was taken as fatal by many researchers. Maudlin's objection relegated TIQM to the sidelines for a decade or so, but in that time a number of authors, including myself, have shown that Maudlin's objection is not at all fatal. While that discussion is beyond the scope of this book, interested readers may consult Kastner (2012, Chapter 5, 2014a) and Marchildon (2006) to see the specifics of those rebuttals to Maudlin's challenge. The basic point is that TIQM is alive and well, and has been elaborated and extended in recent years.

Much of what is contained in this book is based on new research that has been vetted by peer-reviewed journals.

A few notes about the presentation: I've tried to avoid technical and formal language as much as possible. However, the reader may come across phrases such as 'there is no fact of the matter,' which may sound rather formal and unnatural. The reason is that in such cases, it's not just a question of whether one knows something or not; rather, there really may not be anything concrete that we could know, one way or the other. So that is the phrase used to describe a situation in which there really is no concrete fact that could be known, even in principle.

I owe special thanks to some very special people for help and support in writing this book. My sister, Judith A. Skillman, is not only a successful poet but a brilliant writing coach. Her expert editing skills whipped many an awkward passage into much more readable shape. Brad Swoboda offered some insightful suggestions for improved clarity. My mother, Bernice Kastner, provided additional valuable comments. My daughter, Wendy Hagelgans, provided some artwork for figures (in particular, the iceberg of Figures 1.1 and 1.2). And finally, my husband Chuck Hagelgans went over everything with a fine-toothed comb, and insisted on understanding every detail. Without their assistance, the book would certainly have been much less clear. Of course, I am fully responsible for any remaining obscurities or inaccuracies in the presentation.

The central interpretational problem of quantum theory is to answer the difficult question: "What is quantum theory really about?" The answer proposed here is that it is about the unseen, but very real, possibilities that lie beneath the observable world. Less than 150 years ago, nobody believed in atoms because they could not be seen. That point of view seems quaint now. All that remains is to open our minds to the full scope of our unseen reality.

I hope you will enjoy reading this book as much as I have enjoyed writing it.

Contents

Chapter 1

The Tip of the Iceberg

Imagine that you are on the deck of an ocean-going ship, approaching what looks like a small mountain of ice sitting on top of the water:

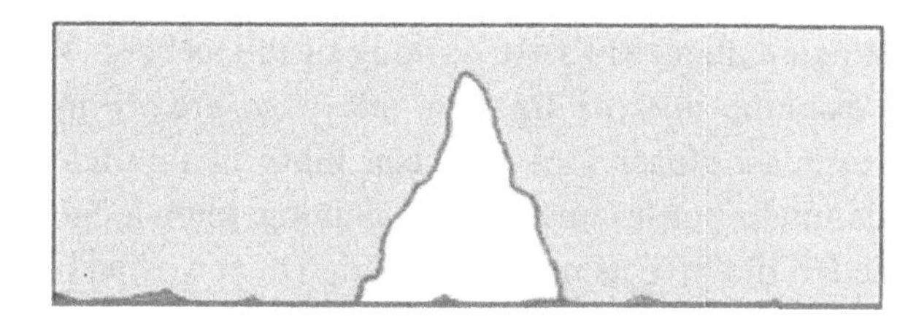

Of course, we all know what this is, but pretend that you had never seen an iceberg before. You get out your telescope and peer through it, and you're able to see this 'mountain' more clearly. You see that it looks confined to a rather small area of water, and that it is indeed made of ice. Curious and intrigued, you decide to approach more closely to get a better look.

Well, knowing the story of the cruise ship *Titanic*, we all know what happens next. That mountain of ice is far more than it appears to be when first observed from the deck of the ship:

One lesson that we can learn from this experience is that things are not always what they seem. Of course, most scientists will tell you that they know this, and that they are trying to discover what things are *really* like, beneath the surface. But what if the surface itself is not what it seems? It turns out that modern physics may be telling us just that. This book will explore the idea that if we think of reality as an 'iceberg,' the older, 'classical' physics describes just the 'tip of the iceberg,' while the new quantum physics is describing the rest of it, beneath the surface of the water, and even the ocean itself.

Let's think again about our first sighting of the iceberg. We used our telescope to examine just the tip. The telescope greatly magnified it, allowing us to get a closer look. We can think of this telescope as a 'classical' (common sense) method of gaining knowledge about the iceberg. Indeed, it did give us more knowledge than we had initially. But that knowledge was only of a superficial kind: the telescope had no way of telling us about the invisible portion of the iceberg beneath the surface. In this sense, our experience is much like that of classical physicists toward the end of the 19th century. At that time, researchers thought that they were in very good shape as far as understanding reality, and that it was just a matter of fine-tuning before they could say that they knew all there was to know about reality by using their classical tools. However, when they began to delve further into the behavior of atoms and electromagnetic radiation, the ship of classical physics ran aground on one of these icebergs, metaphorically speaking. Their classical theories failed to explain to them what was happening. And so, to try to understand this new problem, a whole new kind of physical theory had to be created: quantum physics.

To see what's involved in jumping from classical to quantum physics, let's start with atoms. Classical physicists had started making theories about atoms, which they considered to be the 'basic building blocks' of matter. They thought of atoms as roughly similar to our solar system: a central 'sun' (the nucleus) with the 'planets' (electrons) orbiting around it:

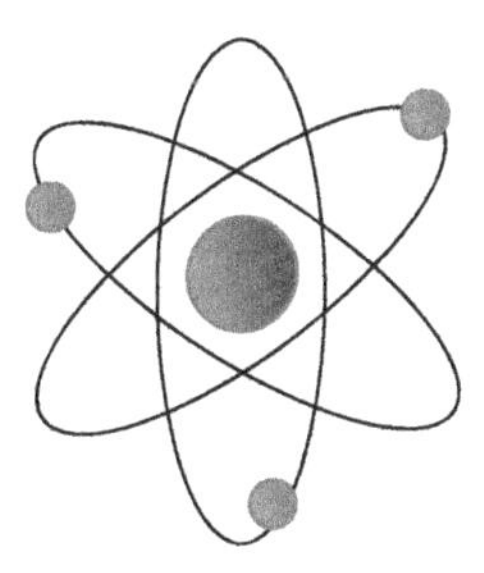

The first time classical physics 'ran aground' was when this solar system model of the atom did not work correctly. One problem was that the electrons had unstable orbits. According to the classical theory, the electrons should gradually lose energy and move in ever-decreasing circles until they crashed into the nucleus; that would bring about a very rapid demise of the 'building block' of matter. The solar system-like model also didn't work well in predicting the kinds of light given off by atoms, which was something physicists could measure. For example, astronomers routinely saw bright lines when they examined the spectra (various wavelengths) of light coming from stars. There was no explanation for these bright lines in the classical theory.

The basic behavior of light presented yet another problem for classical physics, which pictured light as a wave. This model of light worked well for unheated objects, but when a certain configuration of matter was heated, the model predicted crazy things; for instance, that such an object would give off a huge amount of ultraviolet and even x-ray radiation! It essentially stated that you could get very bad sunburn by standing next to a slightly warm oven. This was clearly absurd.

For these reasons, physicists had to go back to the drawing board and develop an entirely new theory of atoms: quantum theory. They were able to formulate a new theory of both atoms and light that gave the right results. That is, the new theory did not state that you should expect to get a sunburn by standing next to a slightly heated oven. It also successfully accounted for the fact that electrons in atoms do not continually lose energy and crash into the nucleus. However, they couldn't state what an atom really looked like with this new theory. The nice, clear, 'solar system' model did not work anymore, and there was nothing to put in its place except for some mathematical formulas; if you used the formulas, you got the right answers. The atom itself seemed to vanish in a puff of smoke:

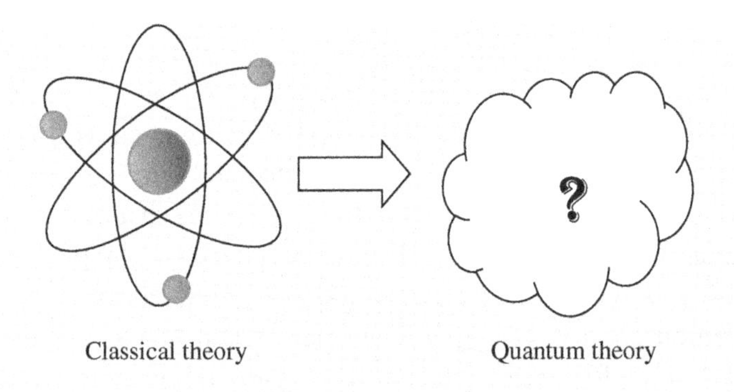

What was this 'smoke'? The answer depended on whom you asked. One of the founders of the new theory, Erwin Schrödinger, said that the quantum 'smoke' was something called a 'wave function,' but aside from saying that it was a solution to a very useful mathematical equation called the 'Schrödinger Equation,' he couldn't tell you what that was. Werner Heisenberg said that the 'smoke' was a pattern of numbers called a 'matrix' (which he also described as a 'laundry list'). Niels Bohr said that you shouldn't even ask him that question, and refused to answer. The remainder of the 20th century consisted of physicists either trying to figure out what the 'smoke' was, or telling each other that Bohr was right. That is, many adopted Bohr's view that the lesson of new physics was that one should not be asking questions about what reality was 'really' like, and just use the 'wave function' or the 'laundry lists' to give results that fit well with what they could observe.

Einstein was one of those who desperately wanted to understand what the 'smoke' really was. He had many debates with Bohr, who was widely regarded as having won those debates. This was because Einstein could never formulate a successful picture of what was underneath the 'smoke.' Bohr was always able to come up with a counterargument that showed that whatever picture Einstein came up with didn't quite work. To those of us who know Einstein as the genius who invented the theory of relativity (as well as many aspects of quantum theory), this is rather surprising. But it shows how seriously quantum theory has challenged our most fundamental assumptions about what reality should be like. And it should also be kept in mind that, even though Bohr was able to refute Einstein's attempted accounts of a realistic picture underneath the 'smoke,' Bohr himself had little to offer besides admonishing us that we shouldn't ask any questions about what reality was like.

This book will argue that we can, in fact, do better by adopting a 'middle way': there is no concrete, spacetime reality corresponding to the 'smoke,' but there is a previously unsuspected, subtle aspect to reality that we never could have guessed at without having been forced into it by the strange behavior of atoms. There is an interesting quote by the late Jeeva Anandan, a particle physicist, on this subject:

> [Quantum] theory is so rich and counterintuitive that it would not have been possible for us, mere mortals, to have dreamt it without the constant guidance provided by experiments. This is a constant reminder to us that nature is much richer than our imagination. (Anandan, 1997)

So what might nature be doing here, underneath the surface of our observable world? We can think of this quantum 'smoke' as representing uncertainty. The famous 'Heisenberg uncertainty principle' describes this aspect of quantum objects such as atoms and their constituents. These objects seem to have an elusive, ephemeral character. In contrast, the old classical physics assumed that everything about an object was concrete and certain. In terms of our 'iceberg' metaphor, everything visible above the water — everything certain and well defined — represents the observable world of space and time. Since classical physics demanded that an object be explained only in terms of what can be well defined within space and time, it could describe only the tip of the iceberg. In terms of this metaphor, Einstein wanted quantum objects to have a clearly-definable place on, or in, the tip of the iceberg. On the other hand, Bohr denied that quantum objects could be found in the visible portion of the iceberg, but also forbade any discussion about what might be underneath the water.

There is some irony in the fact that Bohr was one of the key inventors of quantum theory, even though he inherited the attitude from classical physics that one should not try to discuss anything about reality 'beneath the surface' of space and time. Again, in terms of our metaphor, Bohr invented a theory that correctly told you that you had better stay far away from an iceberg, but he insisted that this theory could never tell you why! In this book, we challenge Bohr's assumption and take a careful look 'beneath the surface,' to see what may really be going on.

Plato's Cave

Before embarking on this journey, let's recall an idea explored by the famous ancient Greek philosopher Plato. Plato made a distinction between the world of appearance, on the one hand, and the underlying reality, on the other. In Plato's thought, the underlying reality may be hidden from us in some way; it may not be directly observable through our usual five senses, but it may exist nevertheless as a vital and very real foundation for the world of appearance that we can directly perceive. He illustrated this idea through his famous allegory of The Cave. Plato's Cave is a story of a group of prisoners who are chained in a dark cave, watching and studying shadows flickering on a wall and thinking that this shadow play comprises everything there is to know about their reality. However, the real objects that give rise to the shadows are behind them, illuminated by a fire which casts their shadows on the wall upon which the prisoners are constrained to gaze. The objects themselves are quite different from the appearances of their shadows (they are richer and more complex). In this allegory, Plato's world of appearance consists of the shadows on the wall, while the underlying reality consists of the objects and the light behind them, both of which give rise to the shadow phenomena that are the only things observable to the prisoners.

Clearly, Plato was arguing that the world of appearance is very limited compared to the underlying reality that gives rise to it. He was saying that the prisoners are mistaken in taking the shadow play on the wall as the full reality simply because it is the only thing they can observe. Moreover, he was suggesting that the world of appearance could be deceiving: the things that we directly perceive may not be what they appear to be.

These sorts of questions about what we should take as 'reality' underlie the exploration in this book. We'll consider in more depth some of the specific philosophical questions about reality in later chapters. For now, it's enough to note that there is an important distinction to be made between appearance (the world of observable phenomena) and reality (what might lie behind that). The latter might be real in a more fundamental sense, even though it cannot necessarily be directly observed through our five senses. In terms of our iceberg, the world of appearance is just the tip of the iceberg; reality is the submerged portion, which is hidden from view but nevertheless has to be taken into account. In the next section, we'll consider how this distinction between appearance and reality ties in with the concepts of space and time usually invoked in the study of physics.

Spacetime

What do we mean by 'spacetime?' This seemingly simple question is actually the gateway to a major controversy among researchers into the nature of physical reality. We'll defer the more controversial aspects and further details for a later chapter, but for now we need to get a basic idea of what is meant by this term in the context of our comparison of classical and quantum physics.

'Spacetime' is a combination of two primitive ideas: 'space' and 'time.' Space pertains to our everyday sense of the distance between ourselves and other objects, and the separation of objects in our field of view. Time is that mysterious quantity counted by the seconds on our clocks (and the candles on our birthday cakes). These two seemingly very distinct concepts are combined into one concept, 'spacetime,' because of Einstein's now well-established theory of relativity. Relativity instructs us that measured quantities such as length and intervals of time are dependent on our state of motion. Despite that, it also says that all observers, regardless of their state of motion, must measure the same speed for a light signal; namely 300,000 km/s. It turns out that in order to take these two facts into account, space and time must not be completely independent quantities, even though common sense seems to tell us they are. Relativity, which has been solidly confirmed by experiment, leads to a picture in which space and time cannot be considered separate concepts, but instead are unified into a single concept: spacetime. We'll consider these ideas more closely in Chapter 7.

For now, the other feature of spacetime that we need to note is that it is the realm of observable phenomena. That is, spacetime corresponds to the world of appearance, as discussed above, in terms of the Plato's Cave allegory. In scientific contexts, the world of appearance, or the world of observable phenomena, is called the empirical realm. Physics is often referred to as an 'empirical science' because it is crucially important for physical theorizing to be well grounded in experiment. Experiment is fundamentally observation, and therefore part of the empirical realm.

Although physical theories make use of mathematics, the field of physics is distinct from the field of mathematics due to the constraint that physical theory must engage, via experiment, with the phenomenal world in order to have any meaningful explanatory value. A theory could have an elegant mathematical formulation, but if its predictions consistently failed to match observed phenomena, it could not be giving a correct account of what's giving rise to those phenomena. Einstein's theory of relativity is an example of an elegant mathematical theory; it has been accepted in part because of its elegance, but mainly because its observational predictions have borne out.

Therefore experiment, which must always take place in the world of appearance, is a crucial 'quality control' on physical theory. Because of this constraint, and because the world of appearance — spacetime — is what is directly accessible to our five senses, it might at first seem natural to assume that spacetime comprises all of physical reality. However, in this book we'll be exploring the idea that this notion is a holdover from classical physics, and that in order to address the riddles raised by quantum theory, we need to be open to the idea that the spacetime arena is not the whole of physical reality.

In the next section, we'll take another look at the idea that reality might consist of 'more than meets the eye' in terms of a classic literary parable, *Flatland*.

Flatland, Spaceland, and... Quantumland?

In *Flatland*, subtitled 'A Romance of Many Dimensions,' Victorian-era author Edwin Abbott entertains a fanciful exploration of higher-dimensional realities. (The story was also a clever and biting social satire of Victorian culture.) In this remarkable and timeless parable, 'ordinary life' is experienced as a two-dimensional world, called 'Flatland.' The exemplar of this ordinary life, and protagonist of the story, is a Square. Abbott creates an entire planar world populated with various polygons. These geometrical inhabitants of Flatland are subject to a hierarchical caste system in which one's socio-economic status increases with the number of sides. Thus, our Square is a humble member of the professional class, while (in the direction of decreasing status) Equilateral Triangles are craftsmen, Isosceles Triangles are soldiers and workmen, and (in accordance with sexist Victorian values) women are just straight lines with no sides at all (at the bottom rung of the social ladder, but also dangerous, since they can pierce a man with their sharp points). In the other direction (of advancing status) are Pentagons, Hexagons, and so on, with a Circle being the highest form of nobility (as its number of sides is infinite).

One day, our Square's peaceful, humdrum, and 'flat' existence is interrupted by the unexpected arrival of a mysterious visitor in his living room. The visitor seems to be a Circle of bizarre properties: he grows and shrinks before the Square's eyes! As the Square relates,

> I began to approach the Stranger with the intention of taking a nearer view and of bidding him be seated: but his appearance struck me dumb and motionless with astonishment. Without the slightest symptoms of angularity he nevertheless varied every instant with gradations of size and brightness scarcely possible for any Figure within the scope of my experience. The thought flashed across me that I might have before me a burglar or cut-throat, some monstrous Irregular Isosceles, who, by feigning the voice of a Circle, had obtained admission somehow into the house, and was now preparing to stab me with his acute angle. (Abbott, 1884)

The strange visitor then announces that he is a Sphere, and that he lives in a 'space' of three dimensions rather than two. (A sphere intersected by a plane looks like a circle; think of cutting an apple in various places and looking at the circular cross-sections.) The Square scoffs at this. In order to persuade the Square that he is not confined to Flatland, the Sphere undertakes a demonstration: he enters a locked cupboard, removes an item, and deposits it somewhere else in the Square's house. Here is

Abbott's account of these events, beginning with remarks from the Sphere:

> [SPHERE:] I have told you I can see from my position in Space the inside of all things that you consider closed. For example, I see in yonder cupboard near which you are standing, several of what you call boxes (but like everything else in Flatland, they have no tops nor bottoms) full of money; I see also two tablets of accounts. I am about to descend into that cupboard and to bring you one of those tablets. I saw you lock the cupboard half an hour ago, and I know you have the key in your possession. But I descend from Space; the doors, you see, remain unmoved. Now I am in the cupboard and am taking the tablet. Now I have it. Now I ascend with it. [SQUARE:] I rushed to the closet and dashed the door open. One of the tablets was gone. With a mocking laugh, the Stranger appeared in the other corner of the room, and at the same time the tablet appeared upon the floor. I took it up. There could be no doubt — it was the missing tablet. [...] I groaned with horror, doubting whether I was not out of my senses; but the Stranger continued: [SPHERE:] Surely you must now see that my explanation, and no other, suits the phenomena. What you call Solid things are really superficial; what you call Space is really nothing but a great Plane. I am in Space, and look down upon the insides of the things of which you only see the outsides. (Abbott, 1884)

Becoming annoyed with the Square's refusal to believe him, the Sphere announces that he can see not only the Square's whole house laid out before him but also the inside of the Square himself. He emphasizes the latter point by poking the Square in his stomach. This segment of the story culminates with the Sphere physically kicking the Square out of his plane and into the world of three dimensions, 'Spaceland,' which finally convinces him. Alas, when the Square returns to Flatland to 'preach the gospel of three dimensions,' he is imprisoned for heresy. In a final irony emphasizing the great difficulty in considering the existence of realities that are not empirically experienced, the Square tries to convince the Sphere that there might be worlds of four or more dimensions, which the Sphere dismisses as utter foolishness.

In the next chapter, we'll begin to examine the riddles and paradoxes presented by quantum theory. The point of invoking the *Flatland* parable in the context of these quantum paradoxes is to suggest that they can be resolved by considering quantum processes as taking place in a realm of more dimensions than can be contained in our usual, empirical reality: the four-dimensional spacetime theater (three spatial dimensions and one temporal dimension). Just as the Square's empirical reality is Flatland, and the activities of the Sphere seemed inexplicable and bizarre from that standpoint, so the empirical realm of spacetime cannot fully encompass the activities of quantum objects that have their existence in a higher-dimensional reality. In what follows, we'll examine these quantum phenomena more closely, and see how they can be more naturally understood by allowing for an analog of 'Spaceland': a high-dimensional realm we might call 'Quantumland.' Recalling the beginning of this chapter, the spacetime realm is just the 'tip of the iceberg,' and the huge portion of the iceberg below the surface lives in 'Quantumland.'

Chapter 2

Quantum Riddles

'I think I can safely say that nobody understands quantum mechanics.'

Richard P. Feynman, Nobel Laureate in Physics

Quantum theory presents us with some very challenging riddles, to which a good interpretation must offer illuminating answers. But first, let us take a step back and consider the nature of riddles in general.

What is a Riddle?

A riddle is a paradox or apparently unanswerable question that does in fact have an appropriate solution. The answer to a good riddle always lies in 'thinking outside the box' in some way; that is, by discarding an inappropriate logical or semantic constraint on our thinking, or by allowing for a new conceptual approach that we had not previously considered. Consider some examples of classic riddles to see how this works:

1. What holds water yet is full of holes?
2. The more you take, the more you leave behind. What are they?
3. A man had a load of wood which was neither straight nor crooked. What kind of wood was it?
   Here are the answers: (1) a sponge; (2) footsteps; (3) sawdust.

Why are these riddles challenging? In trying to answer the first riddle, we think only of containers with one large space surrounded by a single watertight surface, and this category of containers does not include an object with many small holes. In the second riddle, we think of 'taking' in a material sense, while the solution consists of objects that are not material. In the third riddle, we think only of an intact piece of wood. In each case, our baseline assumptions are too restricted, or our set of concepts is not sufficiently diverse, to permit us to arrive at these perfectly natural and appropriate solutions. Part of the humor in a riddle is that we can laugh at ourselves for not thinking more creatively.

In the same way, quantum theory presents us with apparently intractable riddles that can only be satisfactorily answered by broadening our 'conceptual toolbox' to allow for an appropriate and natural solution; one just as unexpected and preconception-shattering as the answers to the above three riddles. Before we get to those quantum riddles, a brief warm-up exercise may be helpful.

Warm-Up: Ideas, Quanta, and Spacetime

Think of a number between 1 and 20, and keep it in mind as if you might be asked to tell somebody what it is. At this point, I could safely say that an idea of some number exists in your mind, even though I couldn't be more specific than that.

If somebody were to ask me 'Where does this idea exist?' I would be unable to provide any more information than 'in your mind,' since we don't know 'where' your mind is located. The mind is a nonphysical entity, and as such it is not located at any particular place or time; that is, it's not located in spacetime.[1] Nevertheless, you know perfectly well what your idea is about; you can experience it directly in a way that I cannot. So ideas can be said to be intelligible and knowable, even if such knowledge is not acquired through the five senses, and even though an idea can't be located anywhere within spacetime. We can therefore conclude that (1) ideas exist and (2) they are knowable on a subjective, mental level, but (3) they are not spacetime entities.

Now, put your number idea on a mental 'back burner,' where you can retrieve it if necessary. The next step is to select some object in your immediate surroundings. An example would be a book, such as the one shown in Figure 2.1. You can sense that object with all five senses, and if

[1]Some might assert that the mind is nothing more than the brain. There are good reasons to deny this, although that debate is beyond the scope of this book. In any case, the concepts explored by this exercise — mainly the distinction between ideas and spacetime phenomena — are unavailable to someone who assumes that there is no substantive difference between the mind and the brain.

Figure 2.1. A number idea is known only to the thinker. In contrast, a tangible object, such as a book, can be publicly verified.

I were in your vicinity, you would be able to show it to me. Assuming that neither of us suffers from color blindness or some other nonstandard perceptual functioning, we would be able to agree on its observable physical properties and where it is located.

In philosophical terms, the existence and nature of the concrete object you've chosen is publicly verifiable.[2] This means that different people can corroborate their own private, subjective impressions of an object or event in a way that convinces them that the object or event they are discussing exists in spacetime. We can think of spacetime as the public world of appearance. We can conclude that objects such as the one you've just chosen from your surroundings (1) exist, (2) are publicly verifiable, and (3) can be located in spacetime.

Now, pick up your number idea from the back burner. If you were to lie and tell me that it was (say) 5 — when it really was 19 — I would not know the difference. Your chosen number idea is not subject to public corroboration in the way the tangible object is, since nobody except you has access to it. So the two big differences between an idea in someone's mind and a tangible object are that the idea is not publicly verifiable and is not located in spacetime.

At this point, consider a famous quote by quantum theory founder Werner Heisenberg. Heisenberg once commented that a quantum system

[2]The technical term for this concept is 'intersubjectively verifiable.'

could be thought of as a form of 'potentia,' an idea that dates back to the ancient Greek philosopher Aristotle. Heisenberg elaborated on this by describing a quantum system as 'something standing in the middle between the idea of an event and the actual event, a strange kind of physical reality just in the middle between possibility and reality' (Heisenberg, 1962, p. 41).

This little exercise is one way to get a feel for what Heisenberg might have meant by this thing that is 'standing in the middle.' So far, we've considered two kinds of entities: (1) an idea, your chosen number; and (2) an actual object, the concrete object you selected from your immediate surroundings. These two things correspond to (1) an 'idea of an event' and (2) the 'actual event,' referred to in Heisenberg's comment above. So this means that a quantum system has a kind of existence that is somehow 'in the middle' between these two extremes that we've considered so far: it is more concrete than an idea, but less concrete than an actual event. We can be more specific by noting that a quantum system: (1) is publicly verifiable, at least in a limited sense, which we'll be examining further; but (2) it is not located within spacetime.

Later chapters will make more clear the extent to which a quantum system is publicly verifiable, and what this means. However, for now, the basic point is that a quantum system can indeed give rise to publicly-verifiable spacetime events, whereas an idea cannot, at least not directly. For example, the internally-held thought of your chosen number leaves no trace on the spacetime realm of events. You could easily give a false report of what your number idea was, and there would be nothing anyone could do that would reveal its falsehood.[3]

In contrast, quantum systems lead predictably (at least in a probabilistic sense) to concrete spacetime events and are therefore subject to public verification. In fact, this is why quantum theory is a successful theory: much of what it says about quantum systems can be publicly corroborated. We'll be exploring some ways in which this happens later in this chapter. Yet despite the amenability of quantum systems to public verifiability, this book will present the case that quantum systems are not

[3]Perhaps we could give you a polygraph test, but these are notoriously unreliable, and that is why they are generally not admissible in court proceedings.

contained within spacetime. It is this 'in-between-ness' that leads to the proposed interpretation of quantum systems as a new form of possibility which is physically real but which transcends the spacetime realm.

Of course, as we've seen above, this idea of quantum systems as a new form of physical possibility is not a new idea: Heisenberg himself suggested it. Others have explored this idea as well. In his popular book, *Quantum Reality* (1987), Nick Herbert refers to quantum states as representing possibilities. Other authors considering quantum systems as possibilities are Lothar Schafer (1997) and John Polkinghorne (1986). The idea that quantum processes transcend the spacetime realm was even acknowledged by quantum theory pioneer Niels Bohr. Bohr referred to the enigmatic 'quantum jump' as a process 'transcending the frame of space and time' (as quoted in Jammer, 1993, p. 189).

Hopefully, this exercise has allowed you to become acquainted with the conceptual possibility of a middle ground — the quantum possibility — between an intangible idea in the mind and a tangible, concrete object or event in spacetime. We'll need this strange new concept as we consider some specific riddles presented by quantum theory.

Wave/Particle Duality

Let us begin this topic by considering light. Light has long been known to be a form of electromagnetic radiation. What is electromagnetic radiation? It is propagating electric and magnetic fields. By 'propagating' we mean that these fields are traveling from one place to another. To state the same basic idea in a manner more in accordance with the picture that we will be developing here, the fields are transferred from one entity to another.

So what is a field? In basic terms, an electric field can be thought of as 'lines of force' that surround a charged object; you are affected by such a field when your hair stands on end after being rubbed with a balloon. The propagating aspect of the field can be visualized as moving ripples or distortions in these lines of force. The magnetic field of a magnet is what causes it to adhere to a refrigerator. It turns out that electric and magnetic fields are intimately related, and one of the triumphs of classical physics was James Clerk Maxwell's work (in 1865) showing that they are actually two different aspects of a single kind of field, the electromagnetic field. Maxwell discovered that this field can propagate by way of its component electric and magnetic fields, and that this radiation travels at the speed of light.

An electric field that is changing in strength creates a magnetic field, and vice versa. In electromagnetic radiation, the two kinds of fields trade roles back and forth as they give rise to each other over and over again. We end up with an oscillation of electric and magnetic fields, which propagates from one place to another; that is, a kind of wave. As the wave cycles through its trough and crest, it covers a certain distance, called its wavelength. Figure 2.2 shows a wide range of wavelengths of electromagnetic radiation; visible light is only a small window in this range.

When you see a rainbow, you are seeing differing wavelengths of visible light, from slightly longer ones (red) to shorter ones (violet). Beyond the visible violet, there are shorter wavelengths corresponding to ultraviolet, then x-rays. Finally there are gamma rays, which are the shortest possible wavelengths of electromagnetic radiation. In the other direction, beyond the red end of visible light and with progressively longer wavelengths, are infrared, microwave (which we use to heat food in microwave ovens), and radio waves (which can have wavelengths many meters long). We should also note that in addition to a wavelength, waves have a frequency, which (for light) is directly related to the energy they carry. The frequency tells us how many wave crests pass a fixed point in a unit

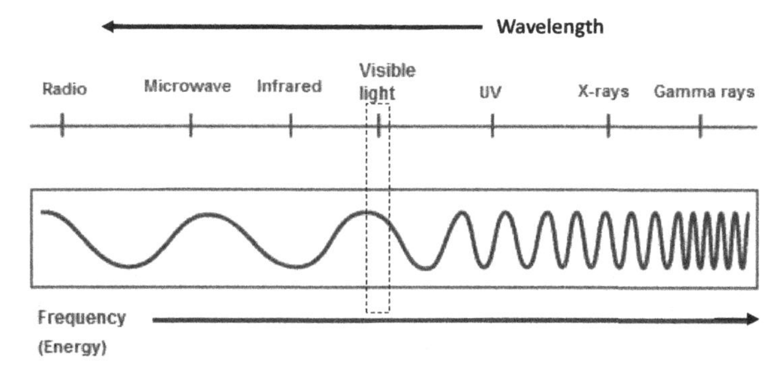

Figure 2.2. The electromagnetic spectrum. The visible light range is indicated by the narrow dashed rectangle.

of time. Waves with shorter wavelengths have higher frequencies and therefore higher energies (see Figure 2.2). For electromagnetic radiation of all kinds, it turns out that when you multiply the wavelength of the wave by its frequency, you get the speed of light: 300,000 km/s.

All waves carry energy. For example, an ocean wave breaking on a beach disturbs the sand on the beach, and that's because it has delivered energy there. We usually think of energy as a quantity that can vary continuously: a tiny ripple in the water delivers a tiny amount of energy, and we know that larger and larger ocean breakers deliver larger and larger energies, all the way up to the enormously destructive energy delivered by a tsunami. However, it turns out that electromagnetic energy can only come in discrete packets or chunks, not continuously, as we would ordinarily assume based on the wave picture. These chunks of energy are called 'photons,' and they behave like particles rather than waves. The fact that light can't be described only by a wave, but also requires a particle-like description, was one of the first discoveries that led to the development of quantum theory. This discovery is known as 'wave/particle duality,' and it is a harbinger of the many riddles that quantum theory presents to us.

The discovery of the particle aspect of light forced a revolution in physics: the quantum revolution. As mentioned in Chapter 1, at the end of the 19th century it seemed very well established that light, and more generally all electromagnetic radiation, was a wave. But as experiments and observations became more sophisticated, the wave theory began to have problems. One problem had to do with shining light on a piece of metal: sometimes this would cause electrons to be ejected out of the metal. This was called the 'photoelectric effect.' However, the energies of the ejected electrons could not be accounted for by assuming that light was a wave, as specified by the classical theory of Maxwell; the energies of the electrons did not match what was predicted by that theory.

According to the wave theory of light, as the intensity of the light shining on the metal increased, the energy of the ejected electrons should increase as well. For waves this would make sense, because the intensity is analogous to the size of the ocean breakers discussed above, and in those terms we could think of the ejected electrons as analogous to the disturbed sand grains on the beach. It seemed like common sense that larger waves should give the electrons a bigger dose of energy. But that didn't happen. Instead, experimenters found that as the intensity of the light increased, more electrons got ejected, but the energy of the individual electrons did not increase. Instead, they found that the energy of the electrons depended on the frequency of the light shining on the metal. This was totally unexpected, and it required a new kind of theory.

Meanwhile, Max Planck was worrying about a different problem. We considered this briefly in Chapter 1: the kind of radiation given off by a certain kind of object called a black body. A black body is an object that absorbs all radiation, of all wavelengths, and does not reflect any of them. For our purposes, we can think of a black body as a black box with a small hole in it that is exposed to incoming electromagnetic radiation of all wavelengths. Radiation can get in through the walls of the box or the hole, but has trouble getting out. Instead, it will just bounce around inside the box for a long time, and the box will continue to heat up. This 'bouncing around inside the box' is modeled in the classical wave picture by standing waves, depicted in Figure 2.3.

These are waves that don't travel in any direction, but simply oscillate up and down in place, like the surface of a drum. However, some radiation does leak out of the hole, and that leaked radiation is a mixture of many different wavelengths. The wave theory of light did a good job accounting for the observed intensity of the longer-wavelength radiation (infrared, microwave, radio waves), but it went off the rails badly at the shorter-wavelength end, with the problem beginning in the ultraviolet range. In fact, it predicted that

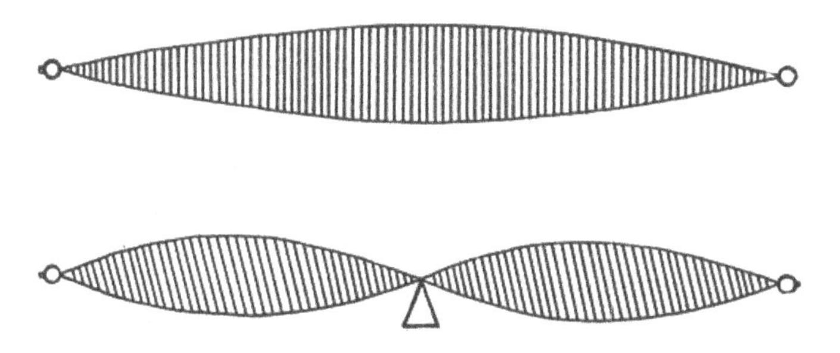

Figure 2.3. Standing waves.

the intensity of the radiation should approach infinity as the wavelengths got smaller and smaller, as if you could get a tsunami of gamma rays out of a small, slightly warm box. This conspicuous failure of the wave theory of black body radiation was known as the 'ultraviolet catastrophe.'

As you might anticipate, both problems — the black body problem and the photoelectric effect problem — were solved by the same idea. This idea was that energy actually comes in finite-sized packets, and is not really a continuous quantity. The general term for these packets is 'quanta,' the plural for 'quantum.' This term comes from the Latin word *quantus*, which means 'how much,' and it also appears in Latin-based English words such as 'quantity.' When Planck reworked his calculations assuming that energy only came in these finite-sized quanta, he found that the 'ultraviolet catastrophe' disappeared.

One crucial feature of Planck's new approach was to assume that the size of the energy quanta depended directly on the frequency of the electromagnetic wave. Specifically, the lower-frequency forms of radiation (such as infrared, microwave, and radio waves) come in smaller-energy quanta, while the higher-frequency ones (such as ultraviolet, x-ray, and gamma rays) come in larger quanta. When Planck assumed that the energy of the black body's standing waves could only come in packets proportional to the frequency of the oscillation, it became harder to create standing waves with those higher energies. In everyday terms, it is as if you were forced to buy larger and more expensive bottles of vitamins as the potency of each dose increases. And this was just what he needed to get the right formula for the black body radiation! This dependence of the sizes of the energy quanta on their frequencies is what cured the 'ultraviolet catastrophe,' by making it harder and harder for energy of higher frequencies to be propagated. In terms of the pharmacy analogy above, you are going to buy fewer bottles of the more potent vitamins if you can only buy them in huge, expensive bottles! But that's apparently how nature operates with the electromagnetic field, and, as it turns out, other kinds of fields as well. Apparently, nature is frugal with her energy.

To visualize the idea of a particle-like aspect to waves, think of a series of sand dunes (Figure 2.4). From a distance, they look like smooth, continuous waves. But as you get closer and closer (this 'zooming in' corresponds to the quantum level), you see that the waves are made up of grains

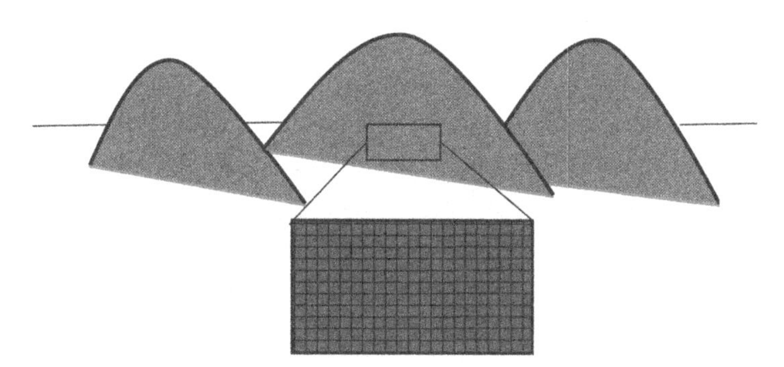

Figure 2.4. Electromagnetic waves are actually composed of chunks of energy, roughly analogous to the sand grains in sand dunes.

of a finite size: the sand crystals. Similarly, energy is generally thought of as a continuous quantity, but as you look closer, you find that it comes in individual packets, or quanta. A photon is a quantum of light, or, more generally, the electromagnetic field. As noted above, waves of shorter wavelengths have higher frequencies, and the higher frequencies mean that the photons corresponding to those waves are more energetic. In terms of our sand dunes, it is as if the dunes with smaller wavelengths (higher frequencies, meaning that the waves are bunched closer together) had larger sand grains. We mentioned gamma waves earlier; these are the highest-frequency, most energetic kinds of electromagnetic waves, and each of their photons packs a real punch.

This dependence of a photon's energy on its frequency is why Planck's idea of quanta also solved the problem of the photoelectric effect. The ejected electrons' energy depended on the frequency of the light hitting the metal because it was the individual photons that kicked out the electrons, and the energy carried by the photons depended on their frequency: the higher the frequency, the more energetic the photons. Increasing the intensity of the light increased the number of photons of a particular frequency; so with more photons hitting the metal, more electrons got ejected. Planck won the Nobel Prize in 1918, and Albert Einstein later won the Nobel Prize, in 1921, for applying Planck's idea to resolve the photoelectric effect. (By then, Einstein had also developed the famous Theory of Relativity, but that was still controversial at the time, so he was officially recognized for the photoelectric effect instead.)

The reader may wonder at this point: if the electromagnetic field is transferred in discrete chunks called quanta, what is it that is oscillating at a particular frequency and with a particular wavelength? Does a quantum 'oscillate'? If so, how? For the past 100 years, physics has been living uncomfortably with this wave/particle duality. Yes, light oscillates like a wave. Yes, it interacts with matter like a particle. But neither aspect alone explains all the ways that light works. The usual approaches to quantum theory also do not provide a good answer to this question; it is one of the paradoxes that the transactional picture can help to solve. In Chapter 3, we'll return to this question, and show how the transactional interpretation (TI) can explain the real physical meaning of wave/particle duality.

Continuing now with our overview, we've considered how quantum theory tells us that things that we thought were waves (i.e., the electromagnetic field and light) also have a particle-like aspect. But conversely, it also tells us that things that we thought were particles — like atoms, and their smaller constituents such as protons, neutrons, and electrons — also have wave-like aspects. This idea was first proposed by Louis de Broglie, and it is part of the mystery of what is oscillating when a photon delivers its energy to another particle. (De Broglie won the Nobel Prize for this idea in 1929.)

This rather mysterious wave nature of quantum objects can manifest itself in various ways. For example, as a wave travels around an obstruction, it bends. This is called diffraction. Figure 2.5 is a sketch of a diffraction effect due to a wave's passage through an opening in a barrier.

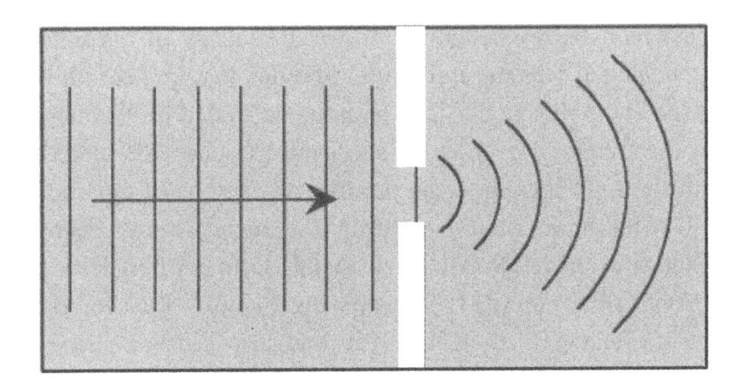

Figure 2.5. Wave diffraction.

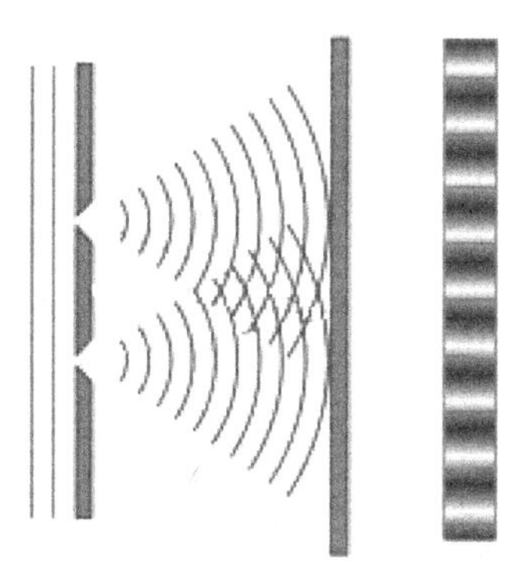

Figure 2.6. A two-slit experiment.

Another feature of wave behavior is interference. It occurs when a wave travels through several openings in a barrier. This is what is seen in the famous 'two slit experiment' with light. As the light waves travel through the slits, the advancing wave fronts from each slit reinforce each other in some regions, and cancel each other out in other regions, yielding a striped pattern at the detection screen (Figure 2.6).

After Louis de Broglie proposed his wave theory of particles, Clinton Davisson and Lester Germer conducted the same experiment with an electron source replacing the light source. Electrons have always been thought of as tiny particles. However, as predicted by de Broglie, the experimenters found that when electrons were sent through a two-slit screen, the detectors displayed the same kind of striped interference pattern. Some researchers supposed that the electrons somehow interfered with each other to produce this interference pattern. But when the electron source was made so weak that only one electron at a time went through the slits, there was still an interference pattern, built up of individual dots where each electron was detected (Figure 2.7). The interference pattern could not be due to the electrons interfering with each other if there was only one electron in the apparatus at a time. Somehow, each individual electron 'interfered with itself.'

It was as if each electron were somehow associated with a wave-like entity that interacted with both slits, and that wave somehow instructed the electrons to be detected more frequently in some places (the bright stripes)

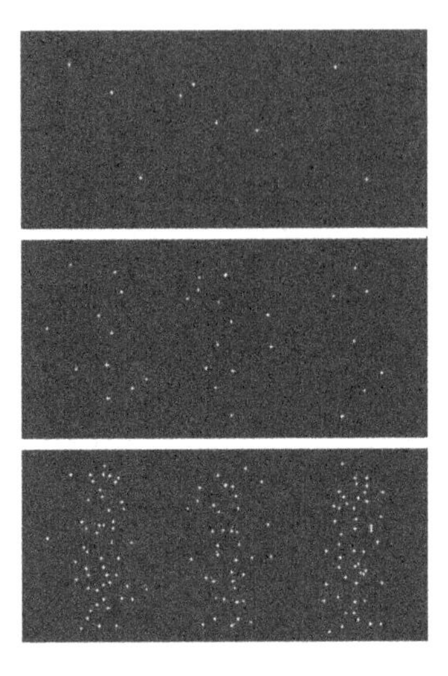

Figure 2.7. A two-slit interference pattern built up, dot by dot, from the detections of individual electrons. There are many vertical stripes; only three are shown.

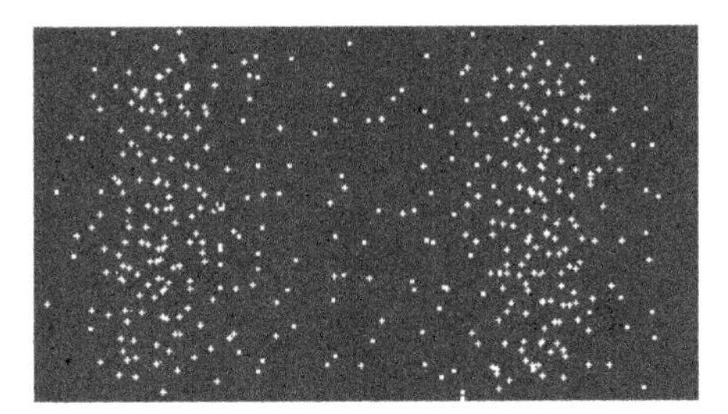

Figure 2.8. What we would see if the electrons only went through one slit or the other.

than in others (the dark stripes). The striped pattern could not be explained by saying that the electron definitely went through one slit or the other; that would yield a completely different pattern, one that did not have the telltale stripes indicating that interference was occurring. Figure 2.8 shows what that would look like.

There would just be two large blobs of electron detections, one corresponding to each slit. In addition, the blobs would be spread out; there is actually a higher probability for electrons to land between the blobs than between the interference stripes of Figure 2.7.

But since we get the repeated interferences stripes rather than the two spread-out blobs (Figure 2.8), the electron cannot be viewed as an ordinary macroscopic object — like a bullet — that goes through only one slit or the other. Instead, it somehow interacts with both slits as it goes from the source to the detection screen. Somehow, an electron is both a wave and a particle; but the wave itself is never directly detected.

Thus, 'wave/particle duality' is a new aspect of the physical world brought to us by quantum physics. In some sense, the fundamental entities of nature can display both wave-like and particle-like features. The crucial insight by Louis de Broglie that even material particles are associated with some sort of ephemeral wave-like oscillation, one that apparently transcends the usual world of observation and experience, is the gateway to the new interpretation explored in this book.

Classical Properties vs. Quantum Possibilities

When a pitcher throws a baseball toward home plate, the baseball has a certain momentum, which is a quantity of motion related to an object's speed. (It would be nice if we represented momentum by the letter 'm,' but that's already being used for 'mass,' so physicists have chosen the letter 'p' to represent momentum.) We can say that its state of motion is 'momentum P,' in that it has the property of 'momentum P' with certainty. Let us represent that sort of classical state by a rectangle (think of this as a nice, solid brick):

P

In contrast, a quantum object is described by a quantum state. This is very different from the 'states' attributed to classical objects such as baseballs. Suppose we have a laser that can emit photons of a certain momentum, p. That quantum mechanical momentum is fundamentally different from the classical momentum of the baseball considered above. The momentum of a quantum object is a kind of *possible* momentum. We can represent the state of this quantum system by a sideways-oriented triangle with a lower case *p*:

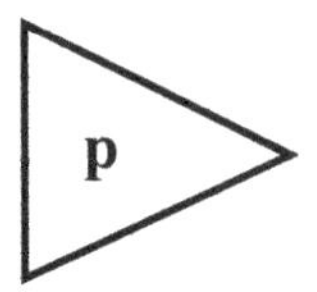

What do we mean by 'possible' momentum? This is the 'standing-in-the-middle' concept presented in the earlier warm-up exercise. The photon represented by this state is a kind of potentiality: a real, yet unobservable, physical foundation for an event that, if it occurred, would be characterized by a particular value of momentum *p*. The event would consist of a photon of momentum *p* (and corresponding energy) being emitted from the laser and detected somewhere, resulting in a loss of energy from the laser and a gain in the same amount of energy by the detector. Yet that event itself is not certain to occur.

This is a very unfamiliar notion, so let us recall Plato's Cave to get an idea of what it could mean. There are prisoners chained in the cave so that they can only see the wall in front of them, which has shadows dancing across it. The prisoners take the shadows as real, because to them they are the only things that are observable. Not observable to the prisoners, but still quite real, are the illuminated objects behind them, which cast their shadows on the wall. Now, suppose the light is turned off. There are no shadows; but those objects, which could cast shadows if the light were turned on, still exist. These objects are the potentialities behind the observable shadows on the wall. In a similar way, the quantum possibilities are necessary precursors to spacetime events, even though they may not necessary result in spacetime events.

Like all analogies, this one isn't perfect. It contains a value judgment: namely that the shadows are impoverished versions of the objects casting the shadows. For our purposes, however, we can just use the following aspect of the analogy: we can compare the shadows on the wall to observable events in spacetime, while the objects casting the shadows represent the quantum possibilities. The basic point is that the quantum possibilities are not contained within spacetime, and are not certain to give rise to specific spacetime events, but they are necessary precursors to any such event. You may be wondering what needs to happen to transform these quantum possibilities into real events that we could experience in space-time. We'll be considering that in detail in the next chapter.

As we will see shortly, these quantum possibilities can 'morph' into one another in a way not allowed in the classical 'brick'-like world. If we think of possibilities as malleable like clay, or evanescent like smoke, this makes sense. Given a lump of clay, it is easily molded into different shapes; given a puff of smoke, it is easily shaped by air currents or other forces. Similarly, quantum possibilities can be molded and reshaped in myriad ways through interactions with our measuring equipment before they can be finally 'cast in stone' and emerge as concrete actualities that could be described by ordinary classical physics.

Although the quantum possibilities represent properties that might be observed, they are not determinate (that is, they are not well defined). Another way of expressing this idea is that quantum properties are not actual, in contrast to those of a baseball or other everyday macroscopic object. Now, in view of this elusiveness, how might we go about measuring a quantum object's properties? In order to do so we need to work with another new concept brought to us courtesy of quantum theory: the notion of an observable.

An observable is basically some aspect of an object that you could measure. In other words, whenever you do a measurement in quantum theory, you are always measuring a quantum system with respect to some particular observable. So, for example, if you decided to measure an object's position, the 'observable' in play would simply be position, usually represented by 'X.' You would expect to find some value in the units of position, which boils down to a distance, such as '5 meters from the wall.' A different observable is 'momentum': if you instead decided to measure the object's momentum, or 'P,' you would expect to find some value in units of momentum instead. We talked about energy earlier; this is yet another observable, which we could label 'E.' One could measure a quantum's energy and get a result in units of energy, such as calories.

Now, all that may seem so obvious that it is not worth stating, since we would expect the same sort of consistency in ordinary classical physics. After all, if we are measuring position, of course we expect to find a particular position. But in classical physics, we can think of such measurements as passive and non-interventionist; as if we can be a 'fly on the wall' that can observe an object without affecting the object in any way. For example, you could measure the length of a baseball bat by setting a measuring tape next to it and reading off the number next to the end of the bat; the bat itself is completely unaffected. So you can think of the bat as having its length independent of whether or not you are observing it.

However, in the quantum realm, a measurement of a particular observable is really a kind of operation performed on a quantum system; it is an active intervention that physically affects the system and often changes it. For example, to measure the position of an electron, you need to bounce a rather energetic photon off it. In doing so, you disrupt the electron; it is then in a different state than it was before the measurement. In view of the intrusive nature of measurement in quantum theory, the measurement of an observable is represented by something called an 'operator,' which can often change the quantum possibility into a different one. For example, we will see later how a quantum possibility that started out labeled by a particular value of momentum can be changed, through measurement of the position observable, into a completely different kind of possibility, labeled by a particular position. This is why quantum theory forces the notion of 'observable' on us; because, despite the superficially-commonplace behavior just discussed, quantum observables behave in rather strange ways. And that strangeness is, in fact, another of the riddles that the theory presents.

What else is strange about these observables? They do not necessarily get along well with one another, as they do in classical physics. In classical physics, you could measure an object's position and momentum in any order you wanted and get consistent, repeatable answers. However, in quantum theory, you cannot do this: you certainly cannot measure a quantum system's position and momentum at the same time, and if you take these two measurements one after the other, the order makes a difference. For example, you will get a different result for position if you measure an object's momentum before measuring its position. That is, the procedure 'First measure X and then P' and the procedure 'First measure P and then X' yield very different results. The theory refers to such pairs (or larger sets) of observables as 'incompatible'. Elementary particle theorist Joseph

Sucher has a colorful way of describing why some observables can be incompatible. He observes that there is a big difference between the following two procedures: (1) opening a window and sticking your head out, and (2) sticking your head out and then opening the window.[4] These are real, physical operations, and clearly something unfortunate is going to happen to your head in one of these procedures but not the other!

This unusual aspect of measurement, and the need to take into account the intrusiveness of quantum observables, is reflected in the famous Heisenberg Uncertainty Principle (HUP). It states that, for a given quantum system, one cannot simultaneously determine physical values for pairs of incompatible observables. Again, 'incompatible' means that the observables cannot be simultaneously measured, and that the results one obtains depend on the order in which they are measured.

However, the HUP is actually something much stronger (and stranger) than the idea that we can't measure both position and momentum because measuring one property disturbs the other one and changes it. Rather, in a fundamental sense, the quantum object, as a possibility, does not have a determinate (that is, a well-defined) value of momentum when its position is detected, and vice versa. For this reason, the HUP is also called the 'indeterminacy principle.' This aspect of quantum theory is built into the very mathematical structure of the theory, which states in precise logical terms that there simply is no 'yes or no' answer to a question about the value of a quantum object's position when you are measuring its momentum. So, returning to our electron being subjected to a position measurement: if the electron was in a state describable by a certain definite momentum, it turns out that you cannot really think of the electron as having had a definite position before you hit it with the photon, because the observables of position and momentum are incompatible. In the state of definite possible momentum, the electron had many possible positions, all at once. Another way of saying this is that the electron was in a superposition of different positions. This is the puzzle of quantum indeterminacy: in general, quantum objects seem not to have precise properties independent of specific measurements which measure those specific

[4]Comment by Prof. Joseph Sucher in a 1993 in a quantum mechanics course at the University of Maryland.

properties. We'll need this notion of a quantum superposition to appreciate the next major quantum riddle.

Schrödinger's Cat

The riddle of Schrödinger's Cat is a thought experiment, devised by its namesake Erwin Schrödinger, to illustrate the so-called 'measurement problem' of quantum theory. Schrödinger knew that the usual approaches to quantum theory had this problem, and he came up with this thought experiment to make it hard to ignore. (Note that this experiment has never been done in the laboratory, and, to my knowledge, no cats have ever been harmed in physics labs.) The problem arises because of the way quantum states are used to label the microscopic objects the theory describes, and the way 'measurement' is usually (and inadequately) handled in the theory. In order to understand the basic problem, we can use the above triangle symbols to play the part of quantum states without having to deal with any mathematics.

First, suppose we have access to a supply of unstable radioactive atoms that we can use in our lab. How does quantum theory describe such objects? An unstable atom has a certain possibility of emitting one or more particles from its nucleus, which is composed of protons and neutrons. An electron can be emitted from a neutron, which then becomes a proton (and that changes the basic identity of the atom). It is impossible to predict exactly when such an electron will be emitted, but the theory provides a way to calculate the average time it will take for this to happen. It does this by considering two possible states for the atom: a state in which the atom has decayed (i.e., sent off an electron) and a state in which it has not yet decayed. We can visualize the 'undecayed' state as in the left-hand picture and the 'decayed' state as in the right-hand picture of Figure 2.9.

The unstable atom can be described by a superposition of these two states. So, using the triangles introduced earlier, let us represent the superposition of these states by a superposition of possibility triangles, labeled U and D for 'undecayed' and 'decayed,' respectively (and we'll put a 'plus' sign in to emphasize the superposition; Figure 2.10).

The story of Schrödinger's Cat begins with this representation of the unstable atom as a superposition of 'undecayed' and 'decayed' quantum

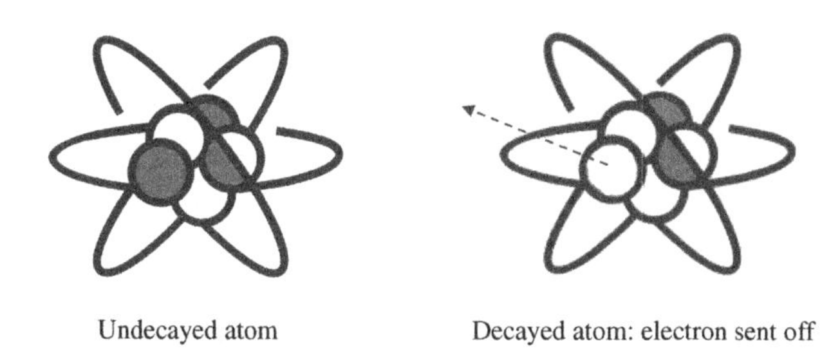

Figure 2.9. Illustrations of the 'undecayed' and 'decayed' atomic state. The emitting neutron becomes a proton, pictured here by changing color from black to white.

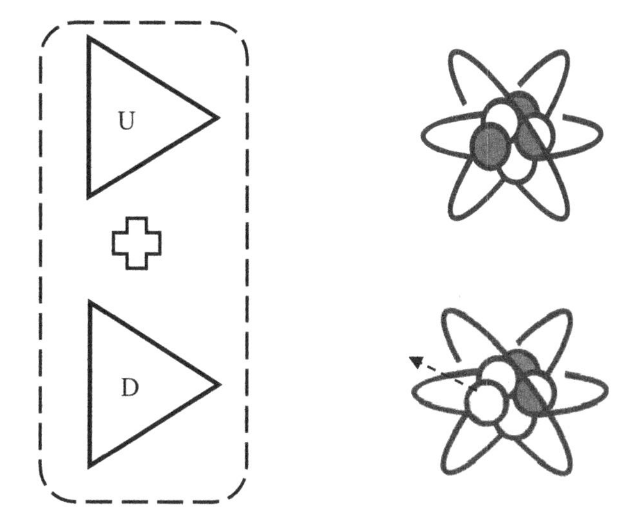

Figure 2.10. The unstable atom as a superposition of undecayed and decayed quantum states.

states. The atom is placed in a box, along with a Geiger counter (which can detect the emitted electron if there is one), a vial of poison gas, and a cat. If the Geiger counter is triggered, this sets off a chain reaction in which the vial is broken and the gas is released, killing the cat (apparently, Schrödinger was not a cat lover). After one hour has passed, the scientist opens the box and looks in to see whether the cat is alive or dead. The 'problem of measurement,' illustrated by this example, is the following: quantum theory (as it is usually understood) treats this scenario by attributing quantum possibilities to all of the interacting parts of the

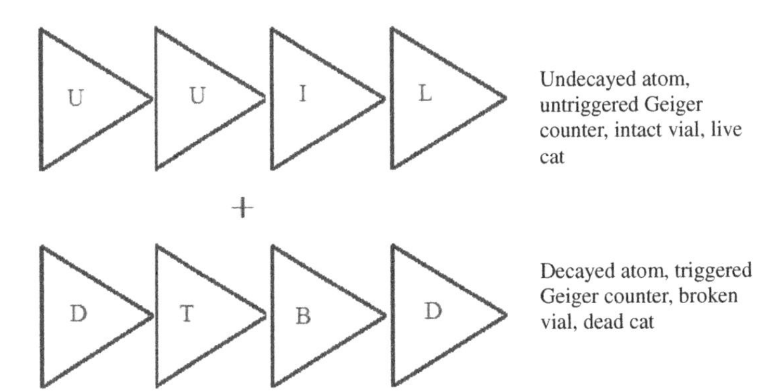

Figure 2.11. The superposition of the atomic states 'infects' the other objects in the experiment.

experiment: the atom, the Geiger counter, the vial, and the cat. That is, it adds a 'train' of quantum states to each of the two basic states of the atom above (undecayed and decayed), so that the states of all the objects in the experiment appear to be described by a huge superposition (Figure 2.11).

The problem now is that the cat and the other macroscopic objects have become 'infected' with the superposition of the genuinely-quantum object (the unstable atom). Each quantum state for the unstable atom has become the 'engine' of a whole train of states labeling the macroscopic objects interacting with it, where now those macroscopic objects seem to have acquired superpositions due to their correlations with the different possible states for the atom.[5] However, we know from everyday observation that we never see Geiger counters or cats in superpositions, so there is something wrong with this description.

Efforts to solve this problem have included invoking the consciousness of an observer in order to 'collapse' the superposition of the macroscopic objects implied by this description. By 'collapse,' we mean to come to some determinate answer; all but one of the possibilities in the superposition go away, leaving one result. That one result is the value we measure.

[5]The interactions consist of forces that result in different states for each of the new interacting objects depending on the state of the quantum. So, for instance, if the atom decays, the Geiger counter experiences a surge of current; if the atom does not decay, there is no surge of current. Since the quantum is in a superposition, the other objects seemingly end up in a superposition as well. This feature of the theory is called 'linearity.'

The consciousness-based approach says that the collapse occurs due to an interaction of the quantum state with the consciousness of an observer. But then we have to decide what constitutes a 'conscious observer,' and even that doesn't answer the question of why or how something unphysical like 'consciousness' would affect such a collapse of a physical object. In the absence of a clear mechanism or process for 'collapse,' there seems to be no way to stop adding quantum states (possibility triangles) for every observer who enters the experiment. That is, if the first observer is Dr. X, the theory seems to require that we add another set of triangles for him, corresponding to the two possible outcomes (Figure 2.12).

Then if another observer, Dr. Y, comes along, we have to add another set for him (Figure 2.13). So there is no end to this train in the usual

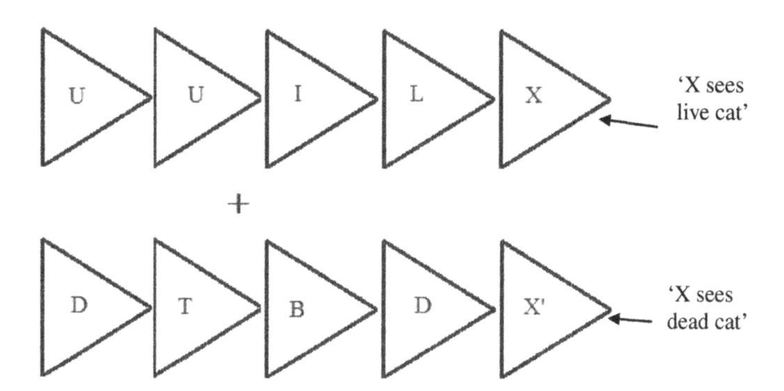

Figure 2.12. Even the human experimenter becomes infected with the superposition.

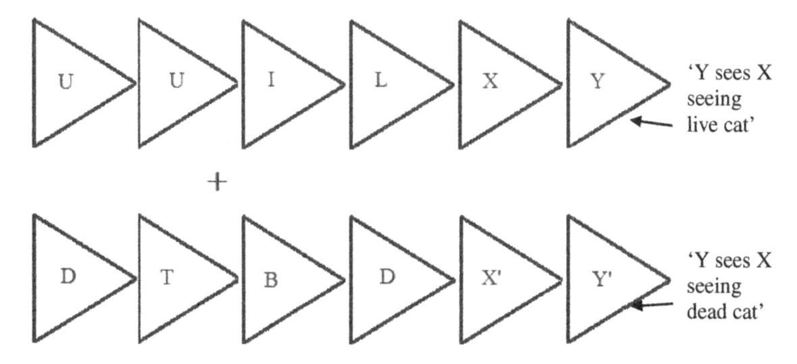

Figure 2.13. The superposition infects everything and everyone involved, seemingly without end (in the usual approach to quantum theory)

method of applying the theory. Another way of putting the problem is that our experience seems to tell us that at some point we have to 'cut' the superpositioned trains off and replace them with something definite, something not in a superposition, which is the outcome we find when we do a 'measurement.' But, in the usual approaches to the theory, the placement of the 'cut' seems to be completely arbitrary, and completely inexplicable. On one side of the cut is the microscopic world that must be described by quantum theory; on the other side of the cut is the macroscopic world that is well described by classical physics. But the standard approach fails to explain why the 'cut' can't be put somewhere in what we know is already the macroscopic world, such as between Dr. X and Dr. Y! This means that what we call a 'measuring device' is arbitrary, too. How do we know what is really 'measuring' something else? It could be the Geiger counter, or an EKG machine that tells us if the cat is dead or alive, or it could be the conscious Dr. X looking into the box, or Dr. Y seeing Dr. X looking into the box. Quantum theory, in its usual formulation, provides no method to specify what a 'measurement' is and why we always see just one result in a measurement when we have started with a quantum object in a superposition. This riddle is the 'problem of measurement.' We will see that the transactional formulation of quantum theory provides a satisfying and illuminating answer to this riddle.

Schrödinger's 'Kittens'

The name attributed to this riddle is due to author John Gribbin, who considered it in great detail in his book *Schrödinger's Kittens and the Search for Reality* (the book was a sequel to Gribbin's *In Search of Schrödinger's Cat*). The 'kittens' are electrons, and the riddle involves an important quantum property called 'spin.' Spin can be thought of as a twirling motion, like that of a figure skater. In Figure 2.14, the 'spin direction' is indicated by the direction in which the skater's head is pointing with respect to the page as she spins counterclockwise. So the spin is 'up' in the left-hand image, and 'down' in the right-hand image.

But usually we are talking about spin under the influence of a magnetic field, and the field can point in any direction you like (these are essentially the 'lines of force' mentioned earlier). Within such a field, quantum

Figure 2.14. A skater illustrating the 'up' and 'down' spin of the electron.

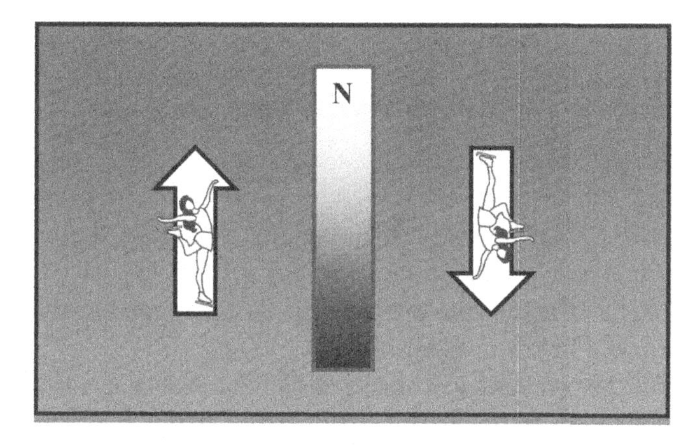

Figure 2.15. 'Up' and 'down' spins relative to a magnetic field.

objects with spin (such as electrons) have only two choices: they can spin in the same direction as the field or in exactly the opposite direction. (If it seems strange that they only have two choices, this is similar to the way in which energy can only come in chunks, or quanta. This is just another way in which the discrete, chunk-like nature of the quantum world appears.) Spin in the same direction as the applied field is called 'up,' and spin in the opposite direction of the field is called 'down.' These are depicted in Figure 2.15. The bar magnet, with its 'north' and 'south' poles, indicates the direction of the magnetic field.

Two physicists, named Otto Stern and Walther Gerlach, designed a device (now called a 'Stern–Gerlach' or SG device) that could measure whether an electron was spinning 'up' or 'down' in a given magnetic field.

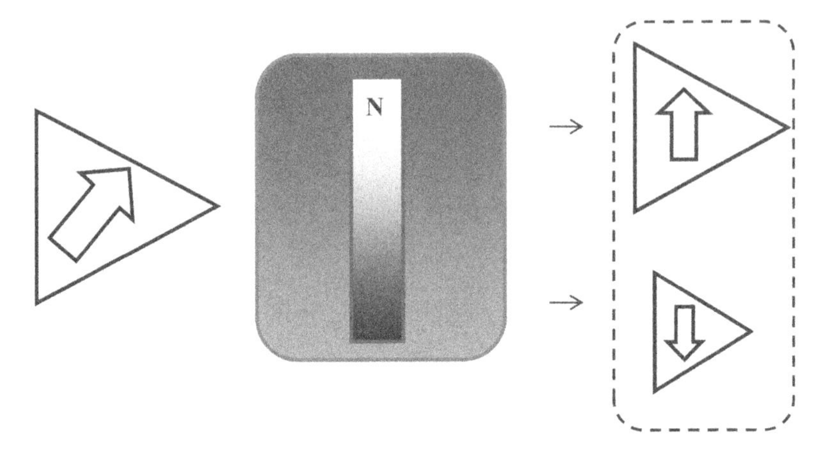

Figure 2.16. A Stern–Gerlach device for measuring electron spin.

The SG device can take a single spin direction possibility that is oriented in any direction and separate it into spatially distinct 'up' or 'down' possibilities along a different spin direction that is determined by the magnet. This is illustrated below in Figure 2.16.

Notice that the original input state gets split into two states with smaller sizes; these sizes are called 'amplitudes'. These output states are 'up' or 'down' along the direction of the S-G magnetic field. If the S-G magnetic field had been aligned in the same exact direction as the incoming state, it would have passed through unaffected and not been split. For an initial state with a different spin orientation than the magnet, the experiment will yield an outcome of either 'up' or 'down,' but each of those outcomes is uncertain; that is, it cannot be predicted with certainty, no matter how well the equipment functions. All we can predict is what the statistics will look like; that is, after a large number of runs of the experiment with the same input state and the same magnet orientation, we can predict roughly how many electrons will turn out spinning 'up' and how many will turn out spinning 'down.'

Note that in the figure the amplitude of the 'up' spin output is larger than the one for 'down' spin. That's because the input state is closer to 'up' than to 'down.' Consequently, it is more likely that the output result will be 'up.' This is similar to playing with a loaded die: a particular number is more likely to come up, but it is not guaranteed. This is the same kind of situation as the two-slit experiment with electrons: we know that more electrons will be found in some parts of the interference pattern than in others, but we cannot predict where any individual electron will land.

Now that we've seen how spin is measured, we can return to the 'Schrödinger's Kittens' riddle. This scenario is based on a thought experiment imagined by Albert Einstein and his colleagues Boris Podolsky and Nathan Rosen back in 1935, which has now actually been undertaken. In the experiment, now known as the 'EPR experiment' in honor of its inventors, we are measuring the spins of two electrons (the 'kittens').[6] These two electrons are entangled, meaning that they share a special kind of two-electron state.[7] One such entangled state is when both electrons have parallel spins, meaning that they have the same orientation with respect to an imposed magnetic field (either both up or both down with respect to that field). However, they are collectively in a superposition as to whether their spins are up or down. So there is no fact of the matter as to whether they are both up or both down, only that they have the same orientation. In terms of our triangles, the state looks like that shown in Figure 2.17.

Since there are two electrons, which could be found in different places, in general you need two separate SG devices to measure their individual spins. The entangled state dictates that if you set the SG devices to measure spin along the same direction, the electrons' spins will always come out the same, either both up or both down.

Now, the riddle is the following: we could prepare these two electrons in this correlated state and send them off to the opposite ends of the galaxy, where two SG devices are placed, ready to measure the spin of each. Suppose that Dr. X and Dr. Y are manning these two devices, and they can freely choose to measure whatever direction they wish. Suppose both Drs. X and Y happen to set their magnetic fields to the northward direction, and Dr. X's electron comes out 'up' with respect to that direction. Then

---

[6]Actually, the version of the EPR experiment discussed here, using spin, was suggested by David Bohm.

[7]This kind of entanglement of two electrons occurs in a helium atom. It can also be generated by a rare form of radioactive decay, in which two electrons are emitted at once from an unstable nucleus.

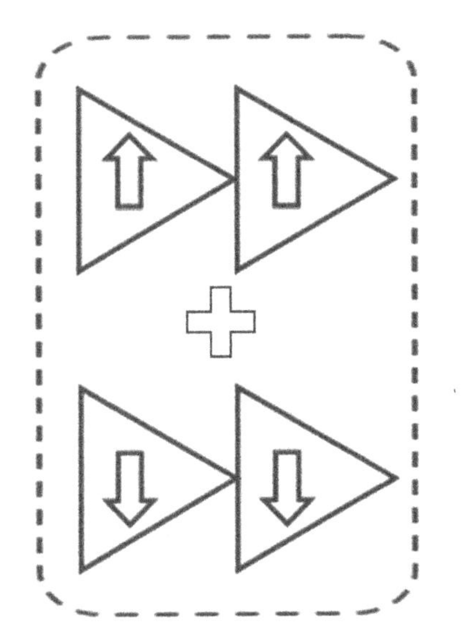

Figure 2.17. A two-electron entangled state with both spins the same.

instantly Dr. Y's electron 'knows' that it must come out 'up' as well; and indeed, Dr. Y will find that result. Now, Einstein's theory of relativity tells us that no signal can travel faster than light, yet there is apparently some sort of communication going on between the two electrons that can instantaneously tell each what is happening to its partner, even across the galaxy! This is what Einstein referred to as 'spooky action at a distance' and is called 'nonlocality' in studies of quantum theory. A 'local influence' is one that propagates at light speed or slower (in accordance with Einstein's theory of relativity), while a 'nonlocal influence' is one that apparently propagates faster than light and therefore seems to be at odds with relativity in some sense. Whatever it is that connects these two electrons does appear to be instantaneously communicated, and is therefore clearly nonlocal. This is one of the paradoxical features of quantum theory, and it is related to the status of quantum objects as possibilities that have their existence beyond spacetime. We'll discuss this in more detail in subsequent chapters.

To summarize, the two big riddles of quantum theory are (1) the measurement problem ('Schrödinger's Cat') and (2) nonlocality ('Schrödinger's Kittens'). This book will discuss how the TI can readily resolve riddle (1) by correcting the mistaken account of measurement that results in an endless 'train' of quantum states. It will address riddle (2) by proposing that nonlocality is indeed real, and that in order to understand it, we must expand our conception of physical reality. In the new understanding, the local realm is spacetime, while quantum objects live outside that realm and thereby escape some of the strictures of relativity. In particular, quantum objects are not subject to locality, which applies only to spacetime objects. We'll discuss TI and its solution to the measurement problem ('Schrödinger's Cat') in the next chapter. To set the stage for untangling the riddle of nonlocality ('Schrödinger's Kittens'), we'll return to the Flatland parable, introduced in the previous chapter.

Strange New Numbers: Emissaries of Quantumland

Consider again the issue of 'wave/particle duality' and the quantum state 'possibility triangle' introduced earlier in this chapter. The quantum state can be thought of as a kind of wave (the de Broglie wave), but it is crucially different from other sorts of waves with which we are familiar. We need to spend a little time considering this aspect of quantum theory, since the TI makes some very interesting and crucial uses of it that differ from all other interpretations of the theory.

An ordinary wave on the surface of the water is described by a mathematical expression that uses only 'real' numbers. What do we mean by 'real' numbers? These are numbers that obey all the usual rules that we were taught in elementary and high school, such as any number (whether positive or negative) multiplied by itself always yields a positive number.

We can represent all the real numbers on a single line, like this:

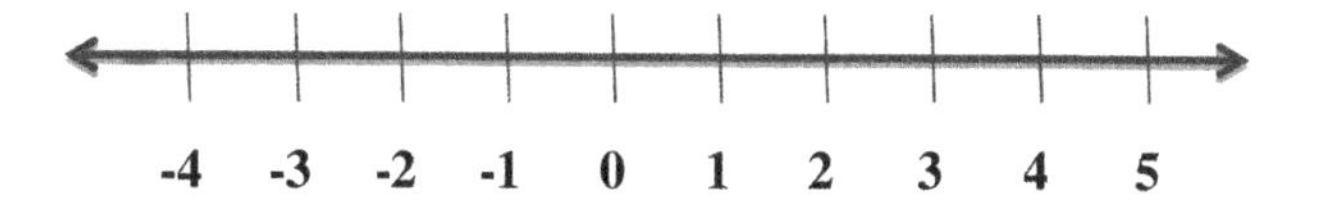

For example, if we take the number three and multiply it by itself (square it), we get nine: in symbols, $3 \times 3 = 9$. Interestingly, we get the same result if we take the number negative three and square it: $-3 \times -3 = 9$. However, if we restrict ourselves to the domain of real numbers, we cannot find an answer to the question 'What number do I square to give me negative nine?' That is, there is no real number that when multiplied by itself will give us a negative number, such as negative nine.

Nevertheless, it turns out that the 'waves' applying to quantum objects require for their description an expanded domain of numbers; a domain that includes numbers that are not 'real' in this sense. That is, in this expanded domain there are numbers that, when squared, give us negative numbers. For historical reasons, such numbers are called 'imaginary.' You can probably understand why the mathematicians who first encountered these numbers called them by the somewhat derogatory term 'imaginary': because they violated the 'common sense' rule that there can be no number that when multiplied by itself yields a negative number. In fact, this is what the great Swiss mathematician Leonhard Euler said about them in 1770:

> All such expressions as $\sqrt{-1}$, $\sqrt{-2}$, etc. are impossible or imaginary numbers, since they represent roots of negative quantities, and of such number we may truly assert that they are neither nothing, nor greater than nothing, nor less than nothing, which necessarily constitutes them imaginary or impossible. (Euler, 1770)

It turns out that these 'imaginary' numbers, together with hybrids of real and imaginary numbers called 'complex' numbers, are needed for describing quantum objects. The fact that we need complex numbers to describe quantum systems is another one of the riddles that quantum theory presents to us. We might call this the 'complexity riddle.' How can we make sense of the fact that physical systems widely acknowledged as being physically real — such as atoms — are described by numbers that aren't 'real'?

To gain some insight into the situation, let us return to our friend, the Square, in Flatland. (Suppose this is at a time prior to the Square's 'conversion' based on his being kicked out of Flatland and seeing the Land of Three Dimensions for himself. This scene is my invention, it is not in the original story.) This evening he's reading the paper in his favorite easy chair in the center of his living room, which is about six meters square. For future reference, we'll label two of the walls of his room according to the directions they face (Figure 2.18). Now suppose that the Sphere decides to look in on the Square from a vantage point above Flatland, so that the Square cannot see him but can only hear him as he says: 'Hello, Mr. Square. How are you this evening? At the moment I'm watching you

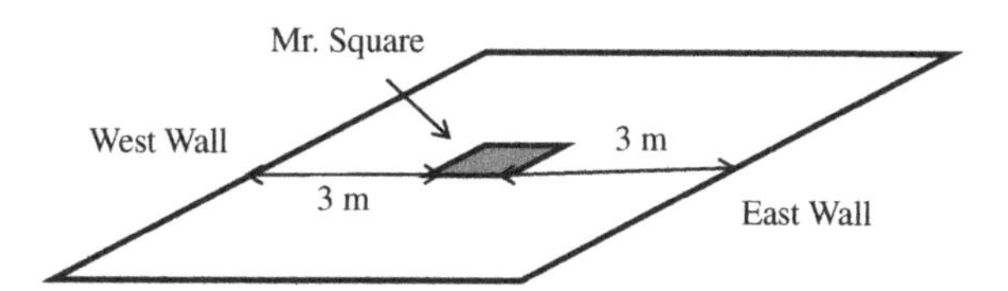

Figure 2.18. Mr. Square in the center of his living room.

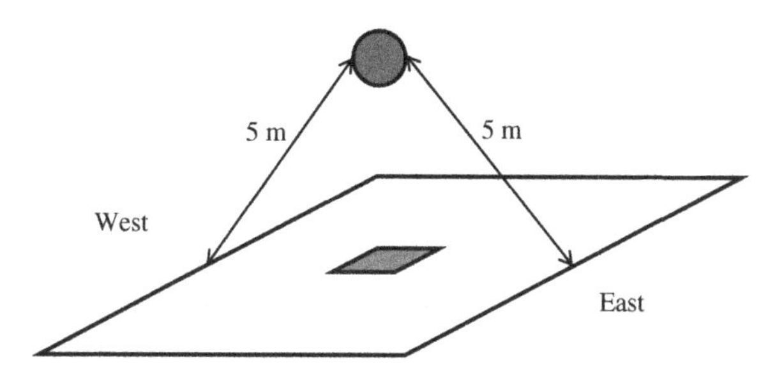

Figure 2.19. The Sphere hovers above Flatland, five meters from both the east and west walls.

from a point that is five meters from both the east and west walls of your living room.'

Well, the Square knows that his living room is only six meters wide, and moreover he is sitting right in the middle of it at a point equidistant from both the east and west walls. So he might say something like: 'That is clearly nonsense. If you were equidistant from both walls, you would be exactly where I am, but you're obviously not here. Moreover, my living room is only six meters wide, but according to your claim, it would have to be ten meters wide. Please stop bothering me with this trickery and foolishness.'

But if we look at the actual situation, we can see that the Sphere is telling the truth about his location (Figure 2.19). To see how this relates to our complexity riddle, we can simplify the situation by looking edge-on, so that the Square's living room is represented by an east–west line. The Square's position is marked by an 'X,' and we can only see his 'south' side:

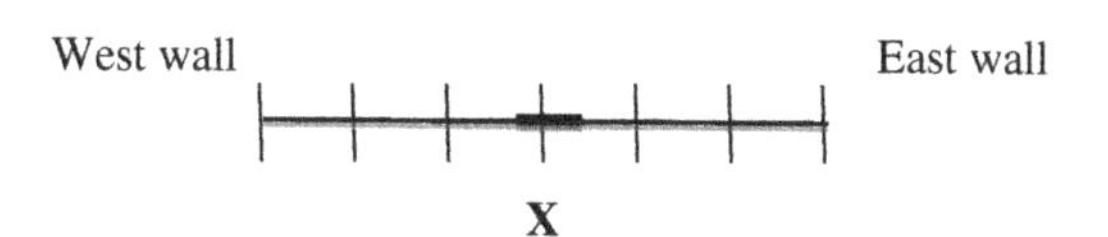

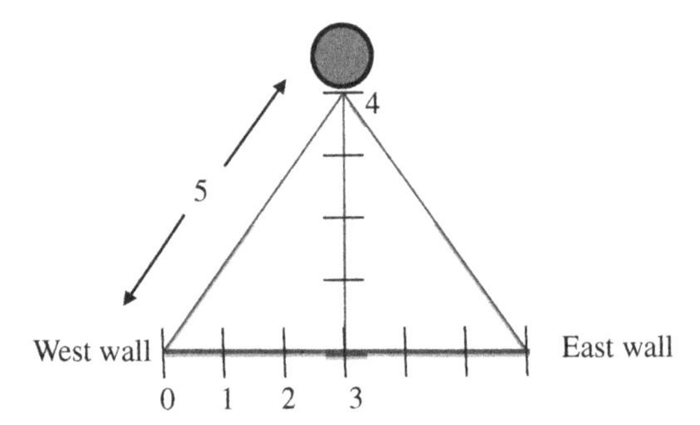

Figure 2.20. The Sphere and the Square, seen edge-on from the south side.

Now we can add a vertical line to represent the Sphere's position above Flatland (Figure 2.20). So if we wanted to describe the Sphere's position, starting from the west wall, we could count three meters to the right, and then four meters up. It turns out that the distance along each of the diagonal lines from the Sphere's position to the walls is exactly five meters. So the Sphere is accurately describing a real physical situation, even though the Square can't make any sense of it from his perspective. That is, the Square can't give a sensible account of the Sphere's existence in terms of the Flatland domain.

Note that this is exactly the same kind of reasoning used by Euler to conclude that the 'imaginary' numbers are 'impossible.' He thought that if the number representing the Sphere's position was not on the horizontal number line, then there must be something wrong with it. But it turns out that imaginary numbers are quite useful: they can help us to describe the Sphere's position, as follows. We can think of the units measuring the distances in Flatland, in this case west to east, as represented by 'real' numbers. The vertical direction represents a new domain that is not contained in the real number line, and therefore has to be labeled by a different breed of numbers. So the units on the vertical axis are labeled by these strange numbers that we call 'imaginary.' In other words, we can think of the Square's living room as part of the (horizontal) number line in Figure 2.20. In more specific terms, if the basic unit in the 'real' or horizontal direction is the number one, then the basic unit in the 'imaginary' or vertical direction is designated i, short for 'imaginary.' It turns out that this number i is

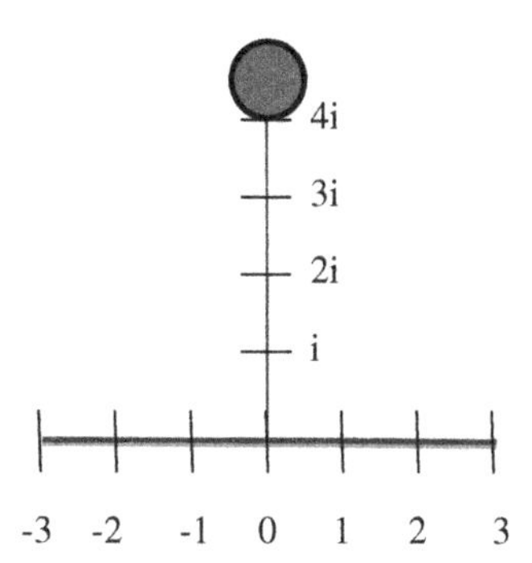

Figure 2.21. The complex plane, illustrated by the Sphere hovering over the Square's living room, seen edge-on.

also the square root of negative one; that is, the number that is multiplied by itself to give negative one.[8]

The procedure described above defines what is called the 'complex plane' by mathematicians. The complex plane is simply all those numbers that have a real component (along the horizontal number line) and an imaginary component (in the vertical direction that can describe the location of the sphere). To gain more insight into this domain, return to the basic diagram of the Sphere hovering above the Square's living room, this time labeled by numbers (Figure 2.21).

Previously we counted from the west wall, but for now we've put the center of our coordinate system right in the middle (where the Square's easy chair is located). Now, using this picture, we can locate the sphere at

[8]Adding a vertical dimension to the number line allows us to represent multiplication by a kind of rotation around the origin (zero). The number i is represented by a 90-degree rotation. We represent $i^2 = -1$ by rotating first by 90 degrees and then rotating again by another 90 degrees, which leaves us at 180 degrees, corresponding to -1. Gauss noted that using language corresponding to this rotational picture would have eliminated a lot of confusion: 'That this subject has hitherto been surrounded by mysterious obscurity is to be attributed largely to an ill adapted notation. If, for example, +1, −1, and the square root of −1 had been called direct, inverse and lateral units, instead of positive, negative and imaginary (or even impossible), such an obscurity would have been out of the question' (Gauss, 1831). But again, this is an example of how one starts with limited concepts that inevitably need to be expanded in order to better understand reality. The number system began for simple counting purposes, for which positive whole numbers were sufficient.

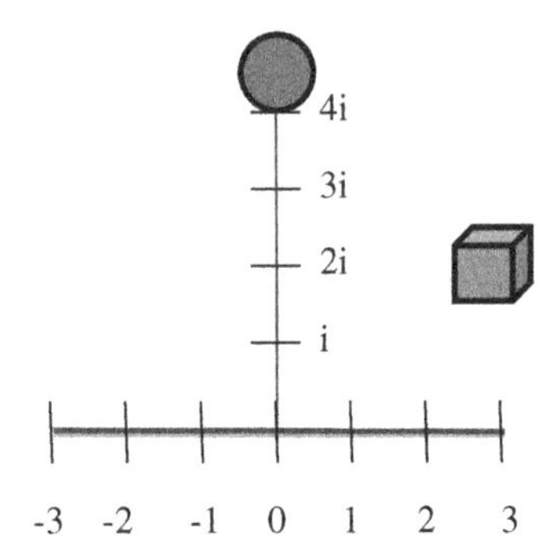

Figure 2.22. The Sphere is joined by another Spaceland friend, the Cube.

the point '*4i*.' But we can locate a point anywhere in the area above the Square's living room by combining real and imaginary numbers, as follows. Consider a friend of the Sphere — perhaps a Cube. The Cube's location is obtained by (starting from the Square's easy chair) going three units to the right and then two units up (Figure 2.22). We indicate this position by the hybrid or 'complex' number $3 + 2i$. We say that such a number has a 'real part,' which in this case is three (the number of units in the horizontal direction) and an 'imaginary part,' which in this case is two (the number of units in the vertical direction).

Why do we care about these complex numbers? Because, to return to the discussion that began this section, quantum objects are wave-like entities that need to be described by complex numbers.[9] That is, we can measure the height of an ocean wave using only real numbers, but the height of the de Broglie waves that describe quantum objects need complex numbers for their description. The height of these waves is called their 'amplitude.' The domain of real numbers is not 'big' enough to describe these amplitudes, in the same way that the Square's two-dimensional world is not 'big' enough for him to describe where the Sphere is located.

[9]Readers familiar with complex notation for real waves might be puzzled by this 'complexity' problem. In the case of real-valued waves, one can represent them by complex expressions for convenience, but in that case the imaginary part has no physical meaning and is ignored. In the case of quantum mechanics, the imaginary part does have physical significance, and cannot be ignored.

Anything observable in the world of appearance must be described by real numbers, so these complex-valued de Broglie waves are not observable. We can now identify (metaphorically speaking) the Square's living room in Flatland as the world of appearance, discussed earlier. Physical reality actually goes beyond that, even though this is hard for him to verify from his vantage point in Flatland. In this metaphor, the Square's living room (and Flatland beyond that) represents our spacetime realm of observational experience. But there is more going on than can be contained in spacetime.[10]

Although the mathematicians who first encountered complex numbers viewed them as merely 'imaginary,' these numbers turned out to play an important role in the physical world. Here is what physicist Freeman Dyson had to say about this interesting historical fact:

> [Mathematicians] had discovered that the theory of functions became far deeper and more powerful when it was extended from real to complex numbers. But they always thought of complex numbers as an artificial construction, invented by human mathematicians as a useful and elegant abstraction from real life. It never entered their heads that this artificial number system that they had invented was in fact the ground on which atoms move. They never imagined that nature had got there first. (Dyson, 2009)

Let us now revisit the idea that quantum systems are subject to public verification, even though they are not spacetime objects and can't be observed directly. Suppose you and a friend decided to conduct a Stern–Gerlach spin-measuring experiment along the following lines. Your friend would prepare electrons in some definite spin orientation, and send them through an SG apparatus. You would be told this preparation state, and would be allowed to choose the direction of the spin measurement by adjusting the SG magnet in any direction you wished. A large number of identically-prepared electrons would be sent through, with the apparatus at the same setting. If your friend were to lie to you about what state he prepared, you would be able to discover his lie in a rather short time,

---

[10]There is even more going on than the expansion of the real numbers to the domain of complex numbers. For more than one particle, the quantum state gains additional dimensions beyond the standard three spatial and one temporal dimensions of spacetime.

based on the measurement results. The simplest scenario in which your friend's lie is revealed is something like this: he actually prepares electrons in the state 'spin up' (in the vertical direction), but falsely tells you that they are 'spin down.' You could detect this lie by setting your SG device for the vertical direction, and of course all the detections would be at the 'spin up' detector, thereby contradicting his statement. So quantum entities have publicly-verifiable consequences; they are not just 'all in your mind.' Yet they are not 'all in spacetime' either. They are something 'in the middle,' as Heisenberg first suggested.

Thus, as Jeeva Anandan (1997) noted in his quote from the previous chapter, nature has a far richer imagination that us mere mortals, and always guides us to an unexpected new picture of reality. Nature has now guided us to the form of a theory that contains strange features that seem to defy the common sense of the spacetime world, just as the Sphere's antics defy the 'common sense' of Flatland. It's natural to conclude that physical reality is simply larger than we thought, and that straightforward and satisfying answers to the quantum riddles can be found within that larger reality. In the following chapters, we'll consider how the transactional picture can resolve the quantum riddles by expanding our understanding of the hidden reality described by quantum theory.

Chapter 3

The Transactional Interpretation: A Conceptual Introduction

'"...paradox" is only a conflict between reality and your feeling of what reality "ought to be".'

Richard P. Feynman, Nobel Laureate in Physics

The original transactional interpretation (TI) of quantum theory was created by Professor John G. Cramer (1986). Dr. Cramer's inspiration for the 'transactional' picture came from the Wheeler–Feynman theory of electromagnetic radiation. This was a novel approach to classical electromagnetic theory, developed in the 1940s by Richard Feynman — who would later go on to win the Nobel Prize in physics for other important contributions — and his then-mentor, John Wheeler.[1] These two eminent physicists created a theory in which radiation was a two-way process. This 'transactional' process, so named by Cramer because it reminded him of a financial transaction, involves an active response by an absorber. The transactional picture differs from the conventional one-way view of radiation, in which an emitter is considered to be the active donor of energy while an absorber is considered a passive receiver of that energy.

This chapter will introduce the basic components of TI, the offer wave and the confirmation wave, and describe what a 'transaction' is. The offer wave, conveying positive energy and other physical quantities, is what is thought of in ordinary quantum theory as the usual quantum system. The confirmation wave is the response of the absorber, which is the novel idea first proposed in the Wheeler–Feynman theory. The confirmation wave has a very strange mathematical property — specifically, it represents negative energy — and that's why it is not considered part of the standard

[1]The Wheeler–Feynman theory was presented in Wheeler and Feynman (1945, 1949).

approach to quantum theory. Most physicists who have considered the idea dismiss it as 'unphysical.' However, we will see that the existence of confirmation waves is implied in all of the calculations that are needed to obtain empirical predictions from standard quantum theory.

Empirical predictions are specific quantities that we can observe, and thereby test the theory to see whether it works by giving correct predictions of the phenomena. In fact, it is the longstanding failure to recognize that the quantum interactions indeed include these allegedly 'unphysical' negative-energy responses of absorbers that has resulted in the measurement problem, discussed in the previous chapter in terms of 'Schrödinger's Cat.' We'll see, in this chapter and the next, just how real these absorber responses are, and how including them in the process of measurement provides a clear solution to the Schrödinger's Cat riddle.

The Basics

To begin with a simple example, let us see how a quantum system travels from a source, called an emitter, to a destination, called an absorber. Suppose we have a quantum system, such as an electron, in a state of definite momentum, p. Its state is represented, as in the previous chapter, by a triangle:

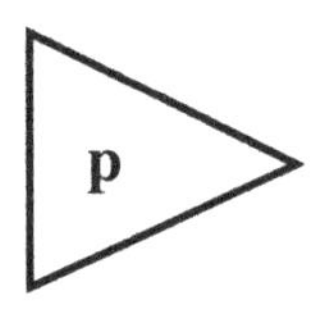

The entities described by quantum states such as this are called *offer waves* in the TI. The use of the word 'wave' acknowledges that the de Broglie wave is a crucial aspect of a quantum system. (Recall from Chapter 2 that even material particles, like electrons, have a wave nature.) The wave is what allows the system to propagate; that is, travel from one entity to another.

Recall that the sizes of the triangles representing quantum states are called 'amplitudes.' Let's examine the idea of an amplitude a little more closely. Those readers familiar with wave motion will recall that 'amplitude' refers to the height of a wave. For example, the small water wave

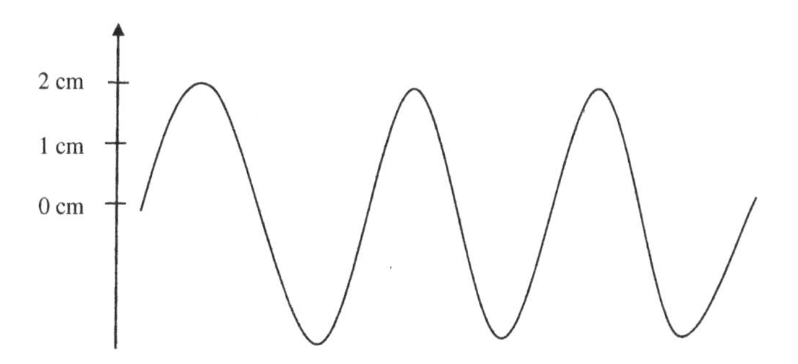

Figure 3.1. A wave with an amplitude of 2 cm.

train pictured in Figure 3.1 has an amplitude of roughly 2 cm. Thus, the amplitude is a measure of how far the wave rises and falls below its starting position with each repetition of the wave motion.

As noted in Chapter 2, it turns out that the de Broglie waves underlying quantum systems have amplitudes that have to be described by complex numbers, like the one used to describe the location of the Cube in the previous chapter. How are we to understand this? What can it mean physically to say that the amplitude of a wave is not necessarily a real number? The set of complex numbers includes the real numbers as a subset, but the basic point is that the quantum amplitude is not restricted to the set of real numbers. It can take on purely imaginary values or combinations of real and imaginary numbers (the combinations being the complex numbers). We'll be discussing this unusual feature of the de Broglie wave in more detail in later chapters, but basically the fact that its amplitude is (in general) not a real number means that it cannot be thought of as contained in the observable, spacetime realm like the ordinary water wave pictured above.

So, to review, TI interprets the standard quantum state, represented by a right-pointing triangle, as an offer wave: it is a physical entity that is offered by an emitter. Now, according to TI, there is another process that needs to be taken into account in addition to the emission of the offer of momentum p above. That process is the absorption of the offer, which gives rise to a *confirmation wave* ('confirmation' for short). We can represent the confirmation by a triangle with the same label, but oriented in the opposite direction; that is, the confirmation is a kind of 'mirror image' of the offer.

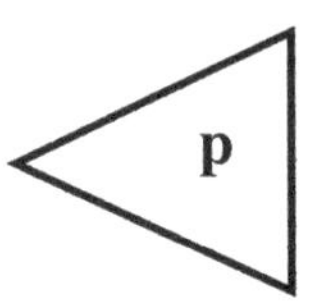

The transactional process can be visualized as a kind of circuit in which the emitting object sends out its offer, which prompts a confirming response from an absorber that travels back to the emitter (see Figure 3.2). This process, an offer responded to by a confirmation, is the basic 'handshake' of the TI. The confirmation is not included in the standard approaches to quantum theory, but it is implied by the mathematical formalism, which includes these backwards triangles.

Like the offer, the confirmation has an amplitude. But here's where things get interesting, and we need to consider again the complex numbers introduced in the previous chapter. As noted earlier, there is something strange about the confirmation: in a rough sense, it conveys negative energy. If the concept of negative energy causes you concern, you are not alone. For years physicists have wrestled with how to deal with negative energy, because it always pops up in solutions of their equations, despite its seemingly 'unphysical' nature. Theoretical physicists are well aware of the fact that you can't really get rid of negative energy. One of Feynman's many contributions to quantum theory was to propose that negative energy represents antimatter. That is indeed one application of negative energy,

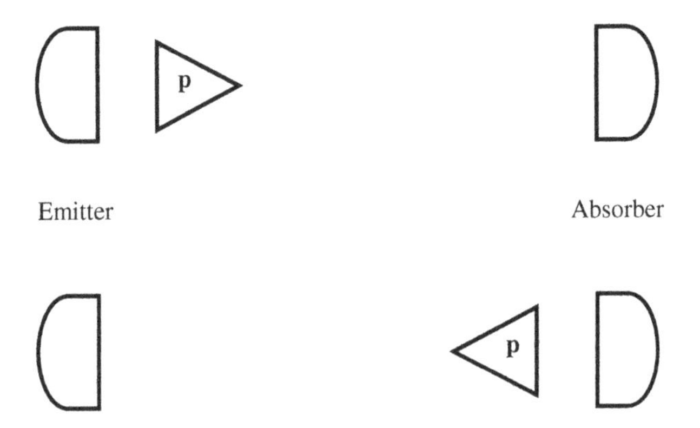

Figure 3.2. The two-step transactional process: (1) an emitter generates an offer; (2) an absorber responds with a confirmation that is a 'mirror image' of the offer it receives.

but negative energy also plays a role in the response of absorbers, which goes beyond the antimatter application.

In mathematical terms, the confirmation is like a 'mirror image' of the offer, and that's why it's indicated here with a left-facing triangle. Picture again the cube of Figure 2.22 and suppose that Flatland itself (the domain of the real numbers) is like a mirror, or pond, that reflects any object on either side of it (Figure 3.3). This 'reflection' creates an object that is the 'mirror image' of the original object, and its position in the coordinate system is described by what is called the *complex conjugate* of the original object's position. As you can see from looking at the picture, this is obtained by changing the sign of the imaginary part of the complex number. So, for example, in the case of our Cube at $3 + 2i$, the 'mirror Cube' is described by the number $3 - 2i$. This complex conjugate is what describes the amplitude of the confirmation. That is, the amplitude of a confirmation is just the complex conjugate, or complex 'mirror image,' of the amplitude of the offer that gives rise to it. It turns out that along with negative energy, this confirmation is directed toward the past, and that is indicated by the opposite direction of the confirmation possibility triangle.

It's important to note at this point that no matter what interpretation you use, you have to go 'through the looking glass' in the above mathematical sense in quantum theory. That is, these 'mirror image' waves are already part of the mathematical machinery of quantum theory. They are always

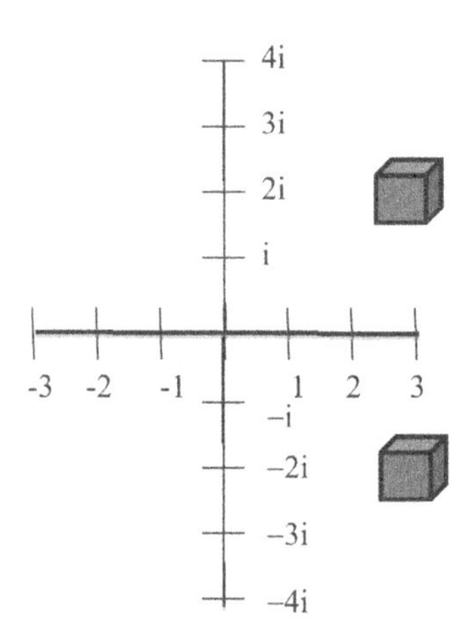

Figure 3.3. The cube's mirror image is its complex conjugate.

part of the standard quantum mechanics calculations needed to use the theory to predict the results of measurements. (We'll take a closer look at the concept behind that calculation below.) However, the usual methods of approaching the theory have no physical account for these complex conjugate expressions, and the calculation has long been just a mathematical recipe with no known physical origin. TI provides a physical origin for the necessary calculation by proposing that the 'mirror image' waves are physical entities that really do exist, and that they are generated in the absorption process.

To review, we have the standard quantum state, represented by our triangle. This is represented in text as **| p >**. TI adds the proposal that when absorption occurs, its 'mirror image' is generated: **< p |**. As noted above, this mirror image wave is already in the mathematical expressions needed to obtain specific predictions from the theory. You may be able to anticipate, based on the symbols, what comes next. Suppose the emitted offer and the answering confirmation are exactly the same. Then, symbolically, what we get is:

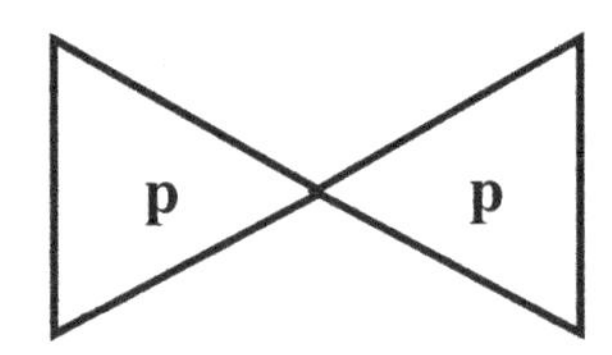

This 'bow tie' is also an expression appearing ubiquitously in standard quantum mechanical calculations.[2] According to TI, this object does not

[2]Congratulations, you've now learned Dirac's 'brac-ket' or bracket notation for quantum mechanics. The right-facing triangle is Dirac's 'ket' and the left-facing triangle is the 'brac'. The 'bracket,' <X|Y>, is an *inner product* in Hilbert space, which is the mathematical domain describing quantum states. The 'bow ties,' |Y><Y| are *projection operators* on that space. TI is the only interpretation that provides a physical basis for a commonly-occurring mathematical expression in the theory: a projection operator multiplied by a squared bracket, e.g., |<X|Y>|^2 |Y><Y|. This quantity represents a weighted incipient transaction in TI. The squared bracket is the probability given by the Born Rule for the detected (absorbed) outcome *Y* given an offer wave with property *X*. Thus, the weight of the incipient transaction is just the probability that it will be actualized. This expression is a great mystery in standard approaches to quantum theory that neglect absorption, while it is self-evident in TI.

yet represent a classical, actualized property p, such as the 'brick' corresponding to a classical object's momentum. However, it is a prerequisite for that property to be actualized. The official name for the physical object represented by the 'bow tie' in the transactional picture is an *incipient transaction*.

How is it, then, that the potential property represented by the 'bow tie' becomes an actual, classical, 'brick'-type property? In the case depicted above, if there is only one offer and a matching confirmation with no competition, then the property p will definitely be actualized (i.e., brought into spacetime) as a classical property. In terms of our symbols, if there are no competing offers and confirmations, then the incipient 'bow tie' transaction $|\mathbf{p}>\,<\mathbf{p}|$ is promoted to a 'brick,' $\boxed{\mathbf{p}}$, through an actualized transaction (Figure 3.4).

To get some insight into how this occurs, recall that the offer and the confirmation are 'mirror images' of each other, and they are in some sense 'waves,' even though they are not literal waves in space and time. We can't think of them as simple water waves or waves on a Slinky; they are more ephemeral than that. A transaction is actualized at least in part by the addition of these two wave-like entities. How does this work? Consider again the Cube and its mirror image, and imagine that we are adding together their positions. Let's do it: we get $(3 + 2i) + (3 - 2i) = 3 + 3 + 2i - 2i = 6$. That is, their imaginary parts cancel, and what we are left with is something entirely 'real': a number that could fit on an ordinary number line. In short, the sum of the offer and the confirmation gives us a mathematically-real-valued wave.

In physics, it's an interesting fact that we seem to find that only mathematically-real-valued entities can be observed in the spacetime realm. So the superposition of the offer and confirmation waves results in a real-valued quantity that can therefore describe our classical, observable rectangle: something that could exist in spacetime. In contrast, the complex-valued

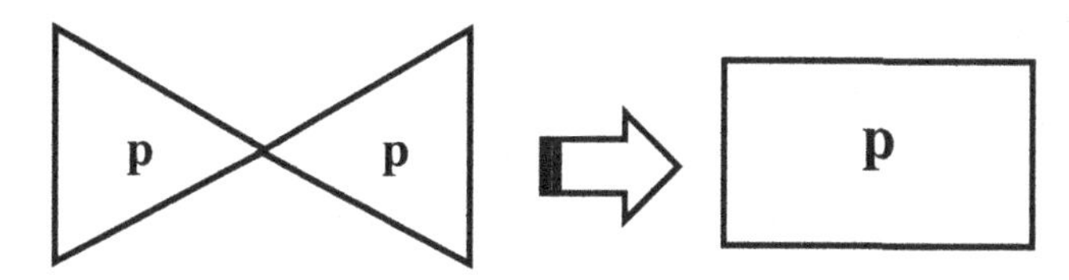

Figure 3.4. An incipient transaction is actualized and becomes an observable event.

triangles (offers and confirmation) are not observable in spacetime, just as the Sphere and the Cube are not observable by the Square in Flatland. Yet they still exist, just as the Sphere and Cube exist.

In the previous chapter, we discussed wave/particle duality. The transactional picture makes sense of this duality by noting that the 'wave' aspect is the possibility offer and its confirmation, while the 'particle' aspect is the actualized transaction that establishes a phenomenon in spacetime, like a detector click or a spot on a photographic plate. Both aspects are necessary components of the process that creates our world of experience.

Throwing the Quantum Dice

What if there are competing incipient transactions? This is where quantum chance enters. In schematic terms for the sake of illustration, let's take a possibility offer expressing a particular pattern from among a variety of possible patterns: stripes, polka dots, etc.:

Suppose there are also other kinds of possibilities involving monochromatic shades of gray. In that case, we would have two kinds of observables: the 'pattern observable' and the 'monochrome observable.' The quantum world allows us to express the 'pattern' offer in terms of those different shades of gray, in much the same way as we can express the number 9 as 2 + 3 + 4.[3] In the quantum world, adding different possibilities means that we have a superposition of those possibilities. So the pattern possibility triangle can just as well be expressed as a superposition of

[3]Of course, the 'shades of gray' do not obviously sum to equal the 'pattern' in the same way as 2 + 3 + 4 = 9. It is actually a type of a vector addition. For more information, see a good introductory quantum mechanics text.

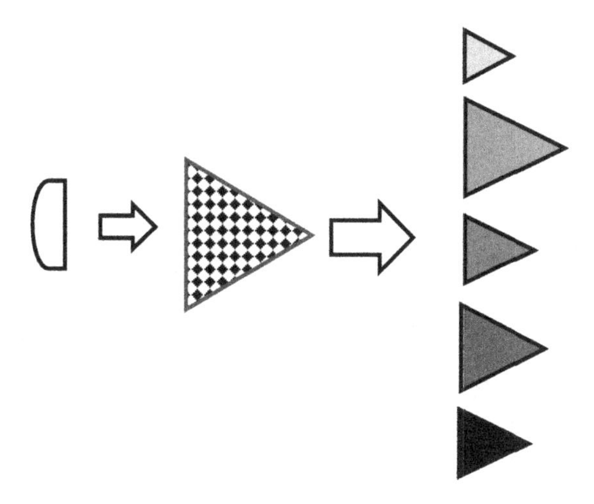

Figure 3.5. A pattern possibility triangle expressed as a superposition of gray triangles of various amplitudes. The gray possibilities are components of the pattern possibility.

offers of different shades of gray, often of (symbolically) different sizes. (In this book, the sizes of the triangles represent amplitudes.[4]) These different shades of gray that need to be added up to correspond to the 'pattern' offer are called *components* of the offer, just as the numbers 2, 3 and 4 can be seen as components of the number 9. The basic idea is illustrated in Figure 3.5.

Now suppose you wanted to measure the shade of gray of the quantum that started out in the given pattern. In order to get an observable result, we would need a measuring device. Such a device would, in general terms, have a needle that points to different possible shades of gray upon completion of the measurement; let's call that the *pointer* of the measuring device. We would set up an interaction between the quantum and our measuring device that correlates its pointer to the different possible shades. Another example of a pointer is simply a detector that fires and thereby tells you that the measurement is complete, and what the outcome was. So the word 'pointer' just means whatever you can look at to indicate to you what the result was. In the case of the Stern–Gerlach (SG) device

[4]In standard quantum theory, the amplitudes of quantum states are indicated by complex numbers multiplying the triangles or 'kets,' which are all the same size.

discussed in the previous chapter, the pointer has two possible readouts: the detectors indicating the different spatial regions to which the 'up' and 'down' offers will be directed.

It's important to note that at the stage in the measurement process in which the interaction between the original offer and the apparatus pointer is established, the process is still completely deterministic, even though we're still dealing with possibilities. That is, in terms of our pattern example above, we can say with certainty what all the possible shades of gray will be, and what their amplitudes are. But only one of those shades of gray will actually be the observed result. Therefore, what is determined so far is not what will actually happen, but rather what the possibilities are, and how potent each one is.

In the presence of absorbers, each of these offers will generate its own matching confirmation. That is, the original offer is transformed into a set of component offers, all of which prompt responses from the absorbers that interact with them. Thus, we get a collection of 'bow ties' of different sizes (Figure 3.6). All of these are incipient transactions, but only one of them can become an actualized transaction that can be viewed a 'classical' property of the 'brick' kind; all the other ones vanish. This is where quantum chance enters. This process is illustrated in Figure 3.7, where the top

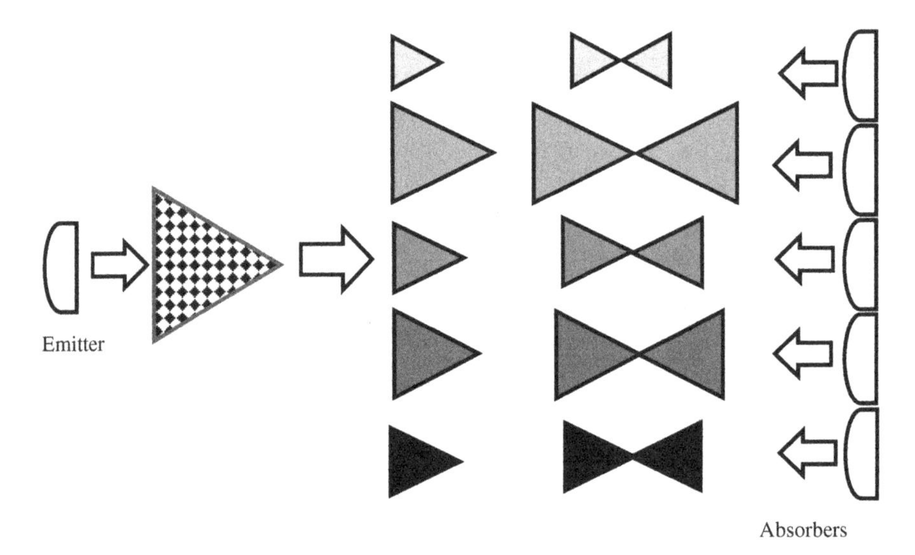

Figure 3.6. A set of incipient transactions arising from the responses of absorbers.

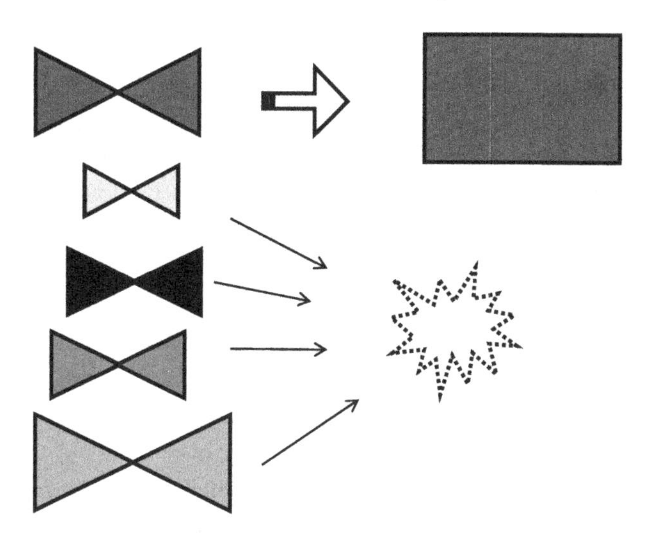

Figure 3.7. One of the incipient transactions is actualized; the others do not result in any observable event.

gray transaction becomes actualized and certain (represented by a brick), while the others disappear. Remember, all the incipient transactions represented possibilities; that is, their essence was merely possible energy rather than real energy. Real energy is only conveyed in the actualized transaction. This is where the 'quantum dice' are thrown: there is no way, even in principle, to predict which of the 'bow ties' will become actualized as an observable 'brick.' However, we do get a hint. The sizes of the bow ties are not merely amplitudes, but rather something more concrete — real-valued probabilities — and these dictate how likely their properties are to be actualized. Recall that the offer waves have complex amplitudes, but once they receive a confirmation response from an absorber, the resulting incipient transaction is represented by a bow tie. The bow tie's mathematical nature is purely real, because the imaginary parts cancel each other out.

In the instance depicted in Figure 3.7, all the other gray-shade bow ties vanish and only the top one becomes an actualized 'event' that could be observed in spacetime. If we did the same experiment over and over, starting with the possibility offer '| **pattern** >' and measuring it with different 'gray scale'-type absorbers, the specific results we'd get would be dictated by the probabilities of the bow ties set up by the responses of absorbers to the various offer wave components. That is, after several runs of the same experiment, there would be more 'bricks' with the color of the larger bow ties than of the smaller ones. If we ran the experiment long enough, we would see that the fraction of each type of outcome is roughly equal to its probability. For example, if the probability of 'light gray' was 20%, and we did 100 runs of the experiment, about 20 of them would have the light gray outcome. The more runs of the experiment we do, the more precisely the fraction of each outcome approaches its probability as predicted by quantum theory.

Why are the probabilities real, 'normal,' non-complex numbers when the offer and confirmation amplitudes are complex numbers with imaginary parts? The short answer is: to get the probabilities, you multiply the amplitudes of the offer and confirmation, and that gives you a real number. We noted above that when you add the offer and confirmation themselves, you get a real wave in spacetime, the strength of which is given by the sum of the real parts (and since their real parts are the same, you just double it). Interestingly, when you multiply the amplitudes of the offer and confirmation, which are mirror images of each other, you also get a real number (the sum of the squares of the real and imaginary parts of the offer wave amplitude).[5] It turns out that that number is always something between zero and one. Expressed in terms of percentages, that number could be anything from 0% to 100%. This is the kind of number that can be interpreted as a probability, just as in games of chance. For example, if you flip a coin, you have a 50% chance — that is, a probability of 0.5 — of getting heads.

Although Einstein famously stated his distaste for this aspect of quantum theory by saying that 'God does not play dice with the Universe,' it does appear that the world has a basic uncertainty or randomness

---

[5]To see this, just multiply the coordinates of our cube example and its mirror cube together: $(3 + 2i)(3 - 2i) = 9 + 3(2i) + 3(-2i) + (2i)(-2i) = 9 + 6i - 6i + 4 = 13$, a real number; again, the imaginary parts cancel each other out. (Note that i times $-i$ is equal to $-(-1) = 1$.) This number is not between zero and one because the position of our cube example involves numbers that are much bigger than the amplitude of a real quantum state. The example just shows how you get a real number when you multiply a complex number by its conjugate.

underlying its apparently solid and deterministic appearances. Whether or not God plays dice, or flips coins, the physical laws themselves do not specify with certainty what will happen, even given full information about (for example) an emitted photon's quantum state. Let's see, using the symbols developed above, how this works in an actual experiment.

The TI Account of Measuring Electron Spin

Let's consider once more the SG measurement used to determine the direction of an electron's spin. Recall that under the influence of a magnetic field, an electron can spin in the same direction as the field: i.e., 'up' or exactly in the opposite direction, 'down.' The direction of the field is indicated by the bar magnet symbol on the SG 'box' (Figure 3.8).

Recall how the box works. It takes an electron offer wave with spin oriented in any direction and splits that into its components of 'up' and 'down,' with respect to the direction of the SG magnetic field indicated by the bar magnet. The 'splitting' of the offer wave occurs because the electromagnetic forces act differently on the up and down components, so they are directed into different spatial locations and detected in different places by different detectors.

In a typical spin measurement experiment, an electron offer encounters the SG magnetic field as depicted in Figure 3.9. The SG device separates the 'up' and 'down' offer components of the electron offer, and these components head off to different detectors. (In this example, we have assumed

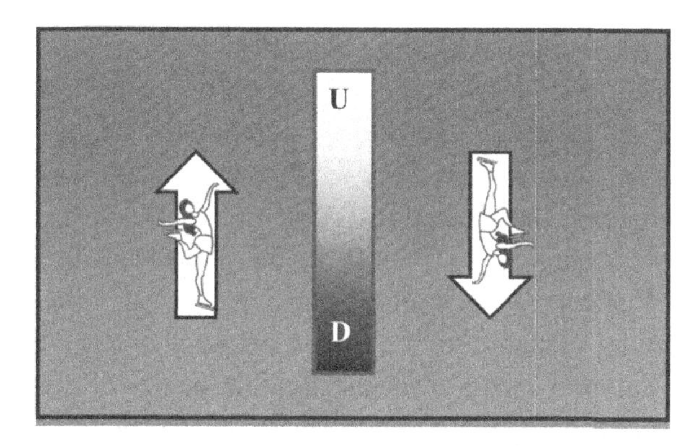

Figure 3.8. The two different spin directions possible in a magnetic field.

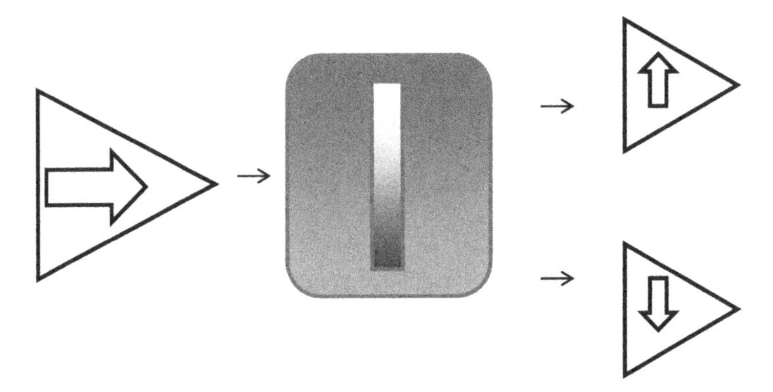

Figure 3.9. The SG device separates this incoming 'sideways' offer into equal components of 'up' and 'down.'

that the input offer is 'sideways,' which is exactly halfway between 'up' and 'down.' The resulting output component offers are then equally split between 'up' and 'down.') The detectors are made up of absorbers that, as noted above, respond with a confirmation that is a kind of 'mirror image' of the offer that they absorbed. It's important to bear in mind that this mirror image is a rather strange object, since it represents negative energy and is directed toward the past rather than toward the future. That may seem 'unphysical' at first glance, but it's important to keep in mind that neither offers nor confirmations are spacetime objects, and neither carries real energy. Again, they only represent possibilities: only one incipient transaction can be actualized, and that is the one that will carry real energy. Let's look at what we have thus far in terms of our symbols (Figure 3.10).

In this schematic illustration, time runs from left to right. However, in a real experiment, the backwards-facing triangles representing confirmations are sent only after the detectors are prompted by the offers. The confirmations are shown in this way so as to make clear where the 'bow tie' representing an incipient transaction comes from: i.e., from the response of an absorber to an emitted offer. Note that both absorbers respond, but only one of them can receive the energy conveyed by the original quantum prepared in the sideways spin state. This happens exactly as in Figure 3.7 with different possible shades of gray, except in this case there are only two possibilities: up or down. Suppose, in this case, it is $|\mathbf{down}\rangle$ that wins the transaction 'lottery' (Figure 3.11). Thus,

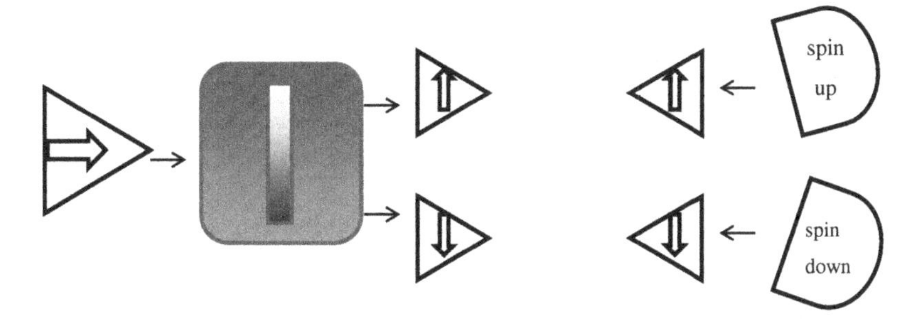

Figure 3.10. Each offer component elicits a confirming response from its detector.

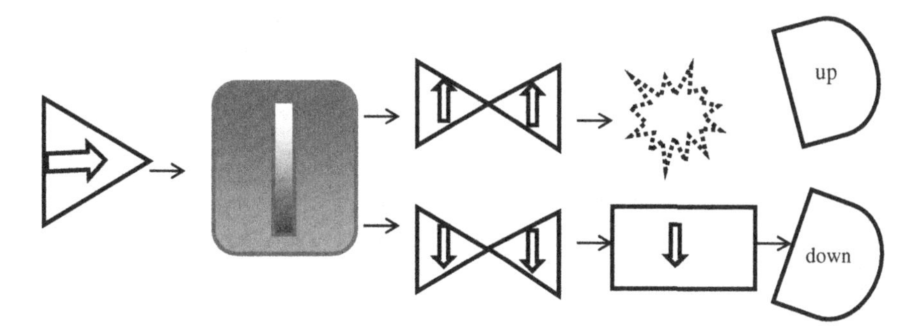

Figure 3.11. The 'down' transaction is actualized.

an observable packet of energy is delivered to the 'down' detector (as indicated by the brick), while no energy goes to the 'up' detector. While we have not shown the emitter of the offer in any of these figures, it's important to note that the transactional process is one that is 'negotiated' between the emitter of the offer and all the absorbers that respond with confirmations.

Before going any further, it's useful to note that we can prepare any electron spin state we like (see Figure 3.12). We can do this by sending any arbitrary electron offer into a preparatory SG device set for our chosen measurement direction and blocking the channel for the outcome we don't want. Meanwhile, we allow the unblocked channel to input the remaining, 'filtered' offer waves into whatever experiment we'd like to perform using the prepared state. So, for example, if we want to prepare the state $|\uparrow\rangle$ with respect to the vertical direction in Figure 3.12, we could block the

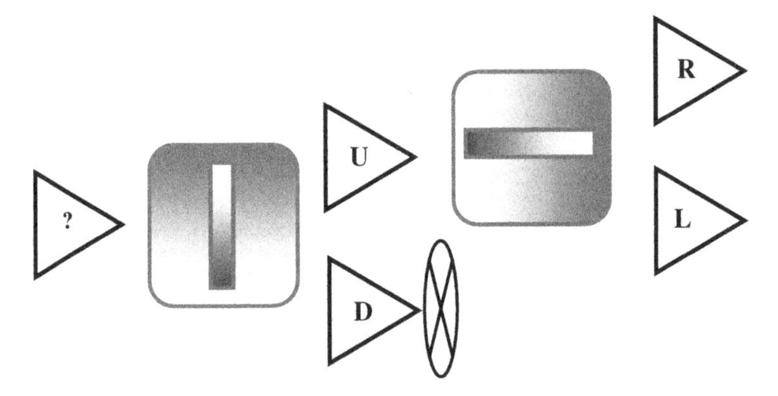

Figure 3.12. Preparing an electron state 'up' in the vertical direction, and sending it through a horizontally-oriented SG box that sorts into 'right' and 'left.'

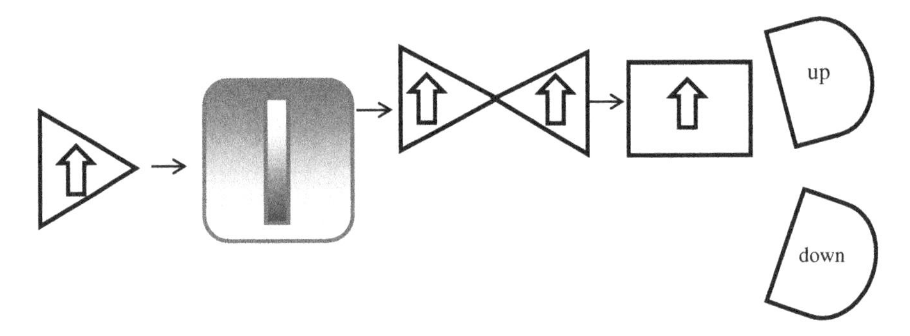

Figure 3.13. An incoming offer wave that is already in the 'up' state.

'down' (south) channel and allow only the 'up' (north) channel as an input into our next device. This way, we would know that only electron offers in the state $|\uparrow>$ would enter our experiment.

Now let us consider the case in which the spin direction of the incoming offer exactly matches the orientation of the SG magnet (Figure 3.13). In this case, there is no offer sent to the 'down' detector, so there can be no absorber response from that detector. We are left with only one incipient transaction, corresponding to the original offer, which will be actualized with certainty. In this way, quantum measurements reliably verify a given initial state, even though the results of measurements of properties other than that of the known initial state are generally unpredictable.

The TI Account of the Two-Slit Experiment

Consider now the two-slit experiment, illustrated in Figure 3.14. We have a source of photons, such as a laser, and in front of the laser is a screen with two narrow slits, labeled *A* and *B*. As the initial offer wave from the laser interacts with the slits, it is transformed into a superposition of offer wave components: i.e., | **A** > + | **B** >. At this point, however, the situation becomes conceptually more complicated than the SG experiment. As these components propagate toward the absorbers in the screen,[6] they change in a way that reflects their likelihood of reaching any particular absorber. The offer wave components reaching the absorber labeled *1* end up reinforcing each other, and *1* sends back a strong confirmation wave which reaches the emitter by way of both slits.

However, the propagation of the offer wave components can also involve changing addition of the components into subtraction. In fact, this is how destructive interference occurs. In Figure 3.14, the two offer wave components reaching point *2* on the screen end up canceling each other

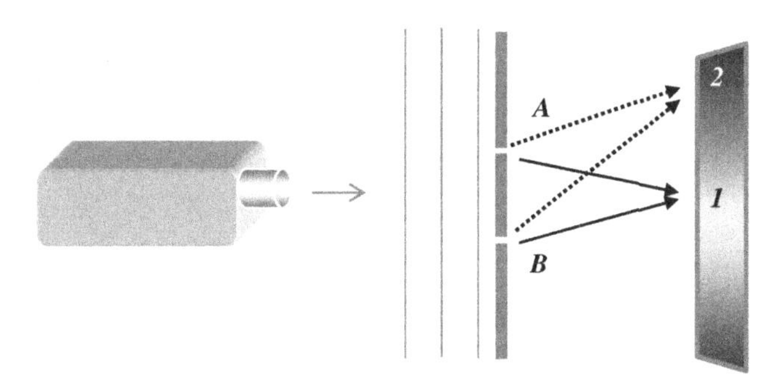

Figure 3.14. The two-slit experiment with interference. The two offer wave components reaching point *1* on the screen reinforce each other, so the absorbers at that point send back a strong confirmation. The confirmation interacts with both slits on its way to the emitter. The two offer wave components reaching point *2* on the screen cancel each other out, and the absorbers at that location receive no offer wave, so they do not generate a confirmation. (In the interest of simplicity, the confirmations are not shown here.)

[6]This 'propagation' is not actually taking place in spacetime, however. More on this point in Chapters 7 and 8.

out, because they transform into something that looks like $|\mathbf{2}\rangle - |\mathbf{2}\rangle$.[7] In this case, absorber *2* receives no offer wave at all, and therefore it does not generate a confirmation.

Meanwhile, other absorbers in the screen receive offer waves of varying amplitudes, from very large to very small. Each of those absorbers sends back a confirmation with a matching (but complex conjugate) amplitude of the offer wave that reached it. This leads to a large set of incipient transactions. In terms of our symbols above, there are a large number of bow ties of varying sizes, one for each absorber in the screen. Only one of these is actualized, and a photon is delivered to the receiving ('winning') absorber. This is the process of 'collapse.' Note that collapse has only occurred at the final screen; both the offer waves and the confirmation waves went through both of the slits. Once again, we see 'wave/particle duality': the waves are setting up a collection of incipient transactions, but only one 'particle' is detected, by the receiving absorber.

Consider now the case in which we wish to observe which slit the photon went through (even though, in the transactional picture, there isn't really a little localized object going through the slit). We can do this using two telescopes focused on each slit (see Figure 3.15).

At this point, we have a similar situation as in the SG experiment described in the previous section. We can think of the top slit, *A*, as the 'up' component and the bottom slit, *B*, as the 'down' component. Each telescope can receive only one of these components, and it responds with

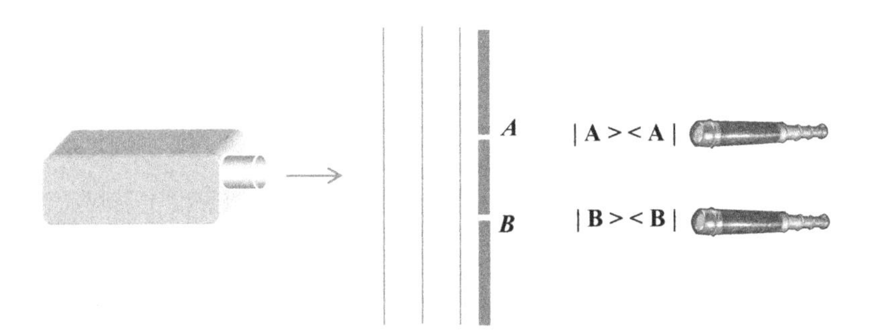

Figure 3.15. The two-slit experiment with a 'which slit' detection.

[7]Obviously, we are glossing over some technical details here. Readers interested in those omitted details are invited to consult Kastner (2012, §4.5).

a matching confirmation. This means, for example, that telescope A sends its confirmation back only through slit *A*, and similarly for telescope B. We end up with only two incipient transactions, $|\mathbf{A}\rangle\langle\mathbf{A}|$ and $|\mathbf{B}\rangle\langle\mathbf{B}|$. When one of these is actualized, either telescope A or telescope B receives a photon of electromagnetic energy. In this case, collapse occurs at the slits, in the sense that the offers and confirmations making up the bow ties only went through one of the slits.

So, to sum up the basic principles of TI: an emitter provides an offer, where the offer conveys physical possibilities (described by amplitudes) for empirical (actualized, observable) events. All absorbers accessible to the emitter respond with confirmations that are mathematical mirror images of whatever offer is absorbed by them. This process results in the 'bow tie' competition; that is, a competition between several potential ('incipient') transactions. Then the 'quantum dice' are thrown, and only one of these is actualized, resulting in a detectable transfer of energy characterized by certain properties (such as 'spin up along the given field direction') to the corresponding detector. All the un-actualized possibilities vanish.

Now, the fact that the quantum state is complex — that is, is described by numbers with real and imaginary parts — makes sense in our picture, since the quantum is acknowledged not to be a spacetime object but rather a carrier of possibility. As such, it does not lives in spacetime (the analog of Flatland in our parable) but in 'Quantumland': a larger, but hidden, realm; just as Spaceland is a larger, but hidden, realm from the point of view of the Square.

What is a 'Measurement Apparatus'?

Here we return to the concern highlighted earlier in this chapter: how to better define concepts like 'measuring apparatus' and 'pointer.' This is a big problem for standard quantum mechanics. Before exploring the TI solution to this riddle, let's see how big a riddle it really is for the standard approach. First, let's recall our discussion of Schrödinger's Cat in the previous chapter. Remember that each time another object is introduced into the picture — be it a Geiger counter, a vial of gas, or even a person — the standard theory seems to require that the object be described by a possibility triangle that gets hitched, like yet another train car, onto an endless freight train. The 'engine' of the train is the original, genuinely quantum object. In fact, some of the other train cars could be genuinely quantum objects, too! But the problem with the standard theory is that it has no way of deciding what is a 'genuinely quantum' object, so it has no way of ending the train.

In the standard approach, when you want to 'decree' that the train has ended, you resort to a calculational rule that works. However, nobody knows why that rule works unless they are using the transactional picture. The rule tells you, in statistical terms only, what you can expect to find when you perform a measurement. The rule, in the standard approach, says that the superposition expressed by the different 'trains' somehow gets transformed into a set of very different objects called 'projection operators.' These projection operators are none other than our bow ties. The rule also says that each projection operator comes with a probability. These probabilities are found by multiplying the amplitude by its complex conjugate (the 'size of the triangle'). Then, that set of 'projection operators' indeterministically (but with the associated probability) collapses to a particular outcome; for example, the outcome corresponding to a live cat. This rule is crucially important, and yet physically inexplicable in the standard theory. It is called the Born Rule, in honor of its discoverer, Max Born, who discovered this rule in 1926 and won the Nobel Prize for this discovery in 1954.[8]

In summary: the Born Rule states that the probability of a particular final outcome, based on a given initial quantum state, is found by multiplying the amplitude of the component of the initial state corresponding to the desired outcome by its complex conjugate. Another way to say this is that it's the absolute square of the amplitude of the component of interest.

Another researcher, John Von Neumann, noticed that the Born Rule arises from that magical 'measurement' process which gives you a set of projection operators, each of which is multiplied by a squared amplitude (that's where the Born Rule comes from). This is the process that was

[8]There have been numerous attempts to 'derive' the Born Rule, but they always appeal to the knowledge of an outside observer or to statistical methods. They don't provide an exact and specific physical process that gives rise to the rule, as TI does.

illustrated in Figure 3.6, in which we had a set of bow ties of different sizes. In that figure, to keep things compact and free of mathematical notation, the sizes of the bow ties represent the squared amplitude, or probability, multiplying each projection operator. So, for example, in terms of our measurement of the grayscale value based on our initial **| pattern >** state, the probability of the outcome 'light gray' is found by squaring the amplitude of the **| light gray >** component that came from the **| pattern >** state, whatever that amplitude happens to be.

At first, Max Born assumed that the required probability was just the amplitude of the relevant final-state component. But then he noticed that the amplitude could be complex, and that it therefore could not give you a legitimate probability value; so he realized that the amplitude must be squared. This is the historical origin of the crucial part of quantum theory that allows it to make predictions about the world of experience: an educated guess.

Standard quantum theory, therefore, has the famous and crucial 'Born Rule' that reliably tells us how to calculate the probabilities of outcomes, but it gives no physical reason for the rule. Moreover, there is no reason, within the standard theory, for why 'measurement' seems to have this special status. That is, the standard theory can provide no account of how measurement somehow transforms the quantum superposition into a set of projection operators, where each is multiplied by a squared amplitude that functions as a probability of the outcome labeled by it. But, as alluded to above, this magical rule may now sound familiar. In fact it is simply the mathematical description of the process described earlier in this chapter: the process in which absorbers respond to offers with confirmations to create a set of 'bow ties.' The 'transformation' described by the Born Rule is simply the transformation of a set of offers (our set of grayscale triangles) into a set of 'bow ties' via the matching absorber responses. The 'bow ties' are represented in the theory by the heretofore mysterious 'projection operators.' The reason for the mysterious squaring operation of the Born Rule is simply the fact that, in the TI picture, the offer generates its own 'mirror image': a matching confirmation the amplitude of which has the same magnitude as the offer that generated it.

You might wonder why the Born Rule calls for multiplying the amplitude by itself (squaring) rather than adding the offer and confirmation together. Here's why: recall that we add the offer and confirmation to get a real quantity of energy, but the summing of the two waves describes the physical nature of the energy delivered from an emitter to an absorber in an actualized transaction. It does not tell us anything about how likely that particular transaction is. The probabilistic aspect comes from the amplitudes ('sizes') of the offer and confirmation waves. The original offer always has the maximum amplitude of one. But all the different components in a superposition obtained from the original offer have been 'shrunk' because they comprise a kind of splitting up of that original offer, much as a deck of cards is split up into smaller hands in a game of bridge. Therefore, each component offer in the superposition has an amplitude that is some fraction of the original one. Its matching confirmation then has the 'mirror image' (complex conjugate) of that same smaller amplitude, as discussed above. The 'bow tie' resulting from the process is a result of a kind of double shrinking, in which the total wave resulting from the circuit 'emitter–absorber–emitter' has been 'shrunk' twice. This process is described mathematically by multiplying the amplitudes together.

So, to return to the question of what a 'measurement apparatus' is: as far as 'collapse' goes, a measurement apparatus is anything that is capable of generating a confirmation in response to an offer; that is, it is an absorber.[9] Absorbers are ubiquitous, which is why it's so hard to retain quantum objects in superpositions. For example, in the Schrödinger's Cat experiment, the Geiger counter generates confirmations in response to the offer corresponding to the unstable atom. This means that the 'train' ends at the Geiger counter with a set of 'bow ties' corresponding to 'undecayed atom' and 'decayed atom.' Now, in the case of an unstable atom, the superposition of its possibility offers will change over time: the amplitude for the 'decayed atom' possibility will increase and the amplitude for the 'undecayed atom' possibility will decrease with time. (The speed of this change is dictated by the specific properties of the atom under consideration; this issue will be examined further in the next chapter.) The 'decayed' bow tie

[9]Of course, a typical measurement apparatus will be more than just an absorber; it will be a system that establishes a correlation with the quantum system, with a pointer that basically corresponds to a set of absorbers. The point here is that an absorber precipitates collapse, and is therefore the most basic measurement apparatus.

(incipient transaction) may be actualized at any time, and an electron ejected, with a probability given by the Born Rule, as discussed above. (The actualizing of the 'undecayed' transaction means that no electron is ejected at that time, and the atom continues to send out its offer.)

The basic point is this: it's not appropriate to ascribe an 'offer triangle' to anything in the Schrödinger's Cat experiment except the atom. The atom is surrounded by absorbers — not just the Geiger counter, but the walls of the box, the cat, anything — accessible to a decay product from the atom. All these absorbers respond with confirmations, which create a set of 'bow ties' or incipient transactions. At this point, one of them is actualized and the quantum superposition is eliminated. Thus, the atom's superposition does not infect the other, macroscopic objects, such as the Geiger counter, the vial of gas, the cat, or the scientists, and this solves the measurement problem. That is, whenever a scientist opens the box, there is already a fact of the matter about whether the atom has decayed or not, simply because of the transactional process, not because he opened the box.

How does Collapse Happen?

At this point in the discussion, hopefully you can see, at least in schematic terms, how the initial offer becomes transformed into a set of incipient transactions (represented in our discussion by bow ties) due to its interaction with a measuring apparatus and its associated absorbers (detectors). Again, we'll examine the details of this in a later chapter. But for now, think of the measuring apparatus as a kind of black box, such as the SG apparatus. The set of bow ties is the basis for the Born Rule, which has no physical explanation in the usual approach. But now: what is it that causes only one of those bow ties to be promoted to a real event, while the others vanish? This is where true 'collapse' occurs, and there is no spacetime story for it; it happens in Quantumland.

We'll examine quantum collapse more fully in the next chapter. But for now, we'll note that this process is much like something called spontaneous symmetry breaking (SSB), which shows up in other areas of physics. Basically, this is a situation in which there exist many possible solutions to a particular equation, but only one can physically occur (see Figure 3.16).

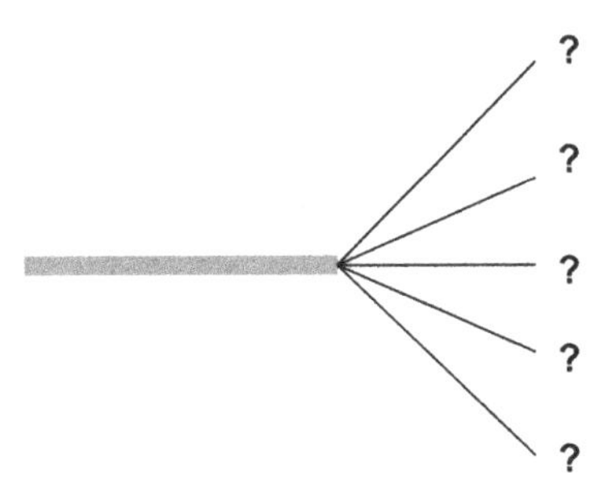

Figure 3.16. Spontaneous symmetry breaking: a theory predicts a multiplicity of states or outcomes, none of which can be 'picked out' by anything in the theory as the realized state or outcome.

Even though there is no causal account for which one is chosen, we still find a particular outcome realized. This is exactly what it means for a process to be truly indeterministic: there is nothing we can point to that 'determines' the outcome. Once again, it appears that, despite Einstein, 'God plays dice.'

A specific example of this SSB phenomenon occurs in the 'Higgs mechanism'[10] in what is termed the 'standard model' of elementary particle theory. (Elementary particle theory studies subatomic particles, such as protons, neutrons, electrons, and other more exotic and short-lived kinds of particles.) According to this widely-accepted model of elementary particles or 'quanta,' pioneered by Steven Weinberg and Abdus Salam, some quanta acquire their mass through a process in which many states are possible, but only one is chosen. However, there is no causal story behind that choice; there is nothing that singles it out from the other possible states. Peter Higgs and François Englert won the Nobel Prize for this discovery (in 2013), which has received enough experimental support to be considered a correct account of something that is really happening in nature. Since the basic concept of SSB is so important, let's review the idea briefly here.

In order to understand the Higgs mechanism, we need to understand a basic aspect of the 'standard model' of elementary particles. This idea

[10]The idea was actually arrived at independently in 1964 by Peter Higgs, Robert Brout, and Francois Englert, as well as Gerald Guralnik, C. R. Hagen, and Tom Kibble.

is that each particle is an excitation of an underlying 'field,' a kind of 'potentiality' akin to the quantum possibilities we've been considering here, but more subtle and basic. You can picture a field as something spread out and very elastic, like a mattress. The undisturbed state of the mattress is called its 'ground' state. If you jump on a mattress, you set it into motion, and those states of motion are called quanta of the field. The harder you jump on it or the heavier the person that jumps on it, the more quanta are produced. As discussed in the previous chapter, these excitations can only occur in discrete 'lumps' or quanta; this was a key discovery by Max Planck, who found that energy always comes in packets of finite size. The size is related to his famous 'quantum of action,' signified by the letter h, and which is now called Planck's constant.

So, these quanta are discrete excitations of an underlying field. That is, they are the energy packets that can be conveyed by that field. The problem facing the standard model is that the theory didn't give these quanta any mass, even though we know, from experiments, that they do have mass. Peter Higgs and, independently, the other researchers mentioned previously discovered that these particles could end up with an effective mass if there was another, background or 'stealth' field that interacted with their own fields. This background field is called the 'Higgs field.'[11] It turns out that the interaction of the field underlying (for example) an electron with the background Higgs field alters the ground state of the electron field in a fundamental way. The theoretical description of this interaction predicts an infinite set of lower-energy ground states (states of lowest energy) for the electron field. However, it is physically impossible for all these ground states to exist; only one gets realized in nature. Even though there is no causal account for the selection of the one Higgs-created ground state that ends up existing, the fact that particles do have mass indicates that one of those infinite number of ground states has been chosen to exist while the others have not. That is, according to the Higgs mechanism, 'many (ground states) are called but few are chosen.'[12]

[11]This field corresponds to the famous 'Higgs boson' that you may have read about.

[12]With apologies to St. Matthew the Apostle (Matthew, 22:14).

The selection of a particular outcome out of a set of equally-viable outcomes seems to run afoul not only of common sense but also of a philosophical doctrine[13] termed 'Curie's principle' (in honor of Pierre Curie, who championed it). Curie's principle states that a particular result (i.e., the choice of one outcome among many equally possible ones) requires a particular cause. The idea behind Curie's principle is that it's not good enough to say that one of the possible outcomes 'just happens'; one must be able to point to a specific reason for that outcome to occur, as opposed to all the others. This principle is illustrated by a humorous paradox, 'Buridan's Ass,' discussed by French philosopher Jean Buridan, in which a hungry donkey is placed between two equally-distant, identical bundles of hay (Figure 3.17).

Figure 3.17. Would a donkey starve to death because he has no specific reason to choose one bundle of hay over another? Nature probably says 'no.' Pictured is a political cartoon (ca. 1900) satirizing U.S. Congress' inability to choose between a canal through Panama or Nicaragua, by reference to Buridan's Ass. (Wiki Open Source; public domain.)

---

[13]Referring to something as a 'philosophical doctrine' simply means that it is presumed to be true on the basis of certain metaphysical or epistemological (knowledge-based) beliefs or principles. Modern physical theory could be taken as indicating that Curie's principle may not be applicable to the physical world, however compelling it may seem to those who have championed it.

According to a version of Curie's principle being satirized by Buridan,[14] the donkey will starve to death because it has no reason to choose one pile of hay over the other. Of course, our 'common sense' tells us that the donkey will find a way to begin eating hay, even though one can provide no reason for it (hence the paradox). Similarly, in SSB, the field in question arrives in a particular ground state, although no specific cause for that choice can be identified. If we take Curie's principle to be applicable to the above, then it appears that nature simply violates the principle (as does a hungry donkey).[15] And indeed, later we'll see how this same basic process is involved in the selection of one of the bow ties (incipient transaction) for actualization.

There is another way of looking at this situation, described by Ian Stewart and Martin Golubitsky (1992). These authors point out that nature seems to be replete with symmetries that are spontaneously 'broken.' One example is a hollow, uniform sphere that is placed under uniform pressure. At some point, it will buckle due to the strain of the pressure on the material holding its shape. When that happens, it cannot retain its spherical form. Instead it will develop a dent somewhere, but where this will happen is inherently uncertain. There is nothing in mathematics describing the buckling that can tell us where the dent will occur. There are many possible final shapes that could result, but the relevant physical theory cannot specify which one wins out.

A famous illustration of symmetry breaking in another kind of physical system appears in the iconic 1957 photo of the splash of a milk droplet by high-speed photography pioneer Harold Edgerton, reproduced in Figure 3.18. Stewart and Golubitsky point out that the pool of milk and the droplet both have circular symmetry, but the 'crown' shape of the splash does not. The

---

[14]Buridan was satirizing the doctrine of moral determinism, which views a person's moral actions and choices as fully determined by past events. In this view, there is always a deterministic reason for why one action is chosen over another. But if there is no difference between the two actions — i.e., nothing which makes one a better choice than the other — then there can be no account for why one is chosen and not the other; so, by Curie's principle, there can be no choice. Hence, Buridan's Ass must starve.

[15]Is there a volitional basis for actualization? Buridan's Ass is hungry, so he chooses to eat one of the piles of hay, even if there is no 'reason' for it. Does nature then express a certain volitional capacity? Or, put another way, could such an uncaused 'choice' be seen as evidence of the creativity of nature?

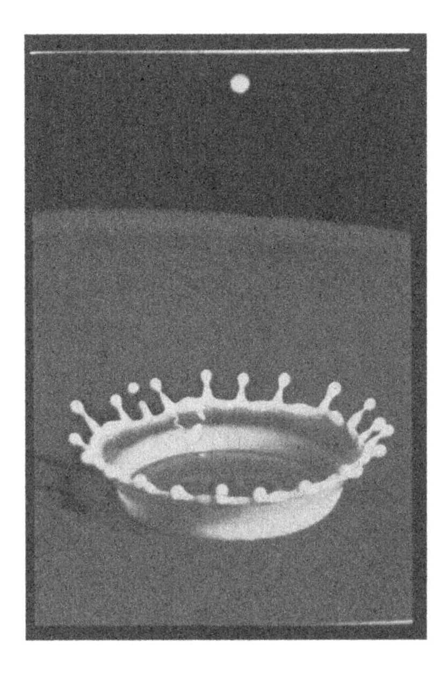

Figure 3.18. Harold E. Edgerton, Milk-Drop Coronet, 1957. © 2010 Massachusetts Institute of Technology. Courtesy of MIT Museum.

crown has a lower degree of symmetry, defined by the 24 points of the crown. By comparison, the circular symmetry has an infinite number of points. This 'symmetry breaking' happens because the ring of milk that rises in the splash reaches an unstable point — a point at which the sheet of liquid cannot become any thinner — and 'buckles' into discrete clumps (the laws of fluid dynamics predict that there are 24 clumps). However, the locations of the clumps are arbitrary; for example, the clump closest to the viewer could just as well have been a few degrees to the left (with all the other clumps being shifted by the same amount). There are an infinite number of such crowns possible, but only one of them is realized in any particular splash.

Thus, Stewart and Golubitsky point out that, while the mathematics describing a particular situation may provide for a large, even infinite, number of possible states for a system to occupy, in the actual world only one of these states can be realized. They put it this way:

> A buckling sphere can't buckle into two shapes at the same time. So, while the full potentiality of possible states retains complete symmetry, what we observe seems to break it. A coin has two symmetrically related sides, but

> when you toss it it has to end up either heads or tails: not both. Flipping the coin breaks its flip symmetry: *the actual breaks the symmetry of the potential.* (Stewart and Golubitsky, 1992, p. 60)

I've italicized the last sentence because it expresses the same deep principle underlying the interpretation presented here: mathematical descriptions of nature, with their high degree of symmetry, in general describe a set of possibilities rather than a specific state of affairs. The authors go on to note that nature can only accommodate one of these possibilities:

> We said that *mathematically* the laws that apply to symmetric systems can sometimes predict not just a single effect, but a whole set of symmetrically related effects. However, Mother Nature has to *choose* which of those effects she wants to implement.
>
> How does she choose?
>
> The answer seems to be: imperfections. Nature is never perfectly symmetric. Nature's circles always have tiny dents and bumps. There are always tiny fluctuations, such as the thermal vibration of molecules. These tiny imperfections load nature's dice in favor of one or the other of the set of possible effects that the mathematics of perfect symmetry considers to be equally possible. (Stewart and Golubitsky, 1992, p. 15)

In the classical case, 'loading the dice' would mean definitely singling out a particular outcome (giving us a deterministic account). That is, in a completely classical world, there would be nothing left to chance, and therefore no 'collapse' of a list of possibilities into one actuality. Rather, there would simply be a smooth progression of the system from one initial state to a single final state. However, in the quantum case, even 'loading the dice' doesn't eliminate the indeterministic aspect; there are still several final states available to the system, with some outcomes (in general) being more likely than others. But it is also possible to have several final states, all of which are exactly equally likely. This is because of that strange quantity, the quantum amplitude, which describes a possibility, not an actual state of affairs. (Recall that we have to multiply the amplitude by its complex conjugate to get a probability; multiplying by the complex conjugate is also called taking the absolute square of a complex number.) The Higgs mechanism, which operates at the quantum level, similarly does not have 'loaded dice' to help with the symmetry breaking needed to obtain particles with nonzero mass.

This section began with the question 'How does collapse happen?' Just as in spontaneous symmetry breaking, collapse in the transactional picture is a genuinely indeterministic process. The most we can say is that the responses of absorbers are the catalysts leading to a set of viable options of which only one can be actualized, since they can't all occur. But the actualization itself is fundamentally indeterministic, so there is no mechanical answer to that part of the question. However, this does not mean that the interpretation has failed to solve the problem of measurement. The most intractable part of the measurement problem is the inability of the usual theory to even define what a 'measurement' is, and why the Born Rule applies to it.

Another aspect of the challenge of defining measurement is to distinguish a genuinely quantum object from a macroscopic object that could be described by classical physics. TI solves this problem by identifying a genuinely quantum object as a possibility offer, while a 'macroscopic' object begins at the point at which a confirmation has been generated. So the basic macroscopic object is a reliable absorber; something that is overwhelmingly likely to generate a confirmation (although not all macroscopic objects are necessarily absorbers). Typical macroscopic objects are systems defined by many actualized transactions. We'll examine this idea in further detail in later chapters. But for now, the basic point is this: in TI, one would never describe a Geiger counter by a quantum state, because a Geiger counter is not just a physical possibility. It is a conglomerate of actualized transactions. But it also retains 'roots' in Quantumland's domain of possibilities because it is comprised of atoms, which can act as emitters or absorbers. The key point is that TI provides a principled method to say why some objects are described by quantum states and others are not.

In the transactional picture, measurement naturally earns its special status. This is because in TI, 'measurement' is simply any process in which confirmations are generated. Since an absorber is an entity that generates confirmations, measurement occurs whenever an absorber is accessible to an emitter. The astute reader will note that we need to be able to say what constitutes an absorber — that is, what is it that gives rise to confirmations? That question will be addressed in the next two chapters.

Chapter 4

Forces and the Relativistic Realm

'A hidden connection is stronger than an obvious one.'

Heraclitus

In this chapter, we'll be discussing the nature of quantum matter and forces, and the interactions between the two. This exploration begins at the relativistic realm, and it will help us to understand how the transactional picture successfully explains the transition between the microscopic and macroscopic realms. The transition between the microscopic and the macroscopic is where our theoretical description transitions from quantum physics to classical physics. This is the point at which the crucial process known as measurement occurs.

Before proceeding, we need to define the relativistic realm. The relativistic realm has two aspects: one involves speeds, the other energies. In general, we are in the relativistic realm when dealing with systems moving at an appreciable fraction of the speed of light, say a few per cent. The relativistic theory is needed at speeds approaching the speed of light, because at these high speeds strange things happen to measurements of time, distance, and mass. The higher the speed, the greater the effect. At lower speeds, there is almost no effect at all, so the relativistic factor can be ignored. On the other hand, when dealing specifically with quantum systems, it's the energy of the system that becomes important. The higher the energy of the quantum system, the more likely it is to give rise to the creation of other particles. The creation of new particles can only be treated by relativistic quantum mechanics. Thus, a generally reliable definition of the relativistic realm is that it deals with situations of high speeds and/or high energies.

It's important, however, to keep in mind that our world is *always* accurately described by relativity, no matter what speeds or energies we're actually dealing with. At lower speeds and energies, relativistic effects are generally negligible and we can therefore use a nonrelativistic theory. But the nonrelativistic theory is always just an approximation. When using it, we need to keep its approximate nature in mind in order not to be misled into thinking that we can disregard all relativistic effects and interactions, when that may not really be the case.

Indeed, one reason the measurement problem has been so intractable for such a long time is that the nonrelativistic theory has been treated as a free-standing theory, one that completely describes all relevant aspects of quantum systems to which it's applied. In fact, even though it can be used to generate useful predictions, it does not completely describe what is going on. Some of the processes it cannot describe are those underlying measurement.[1] These are what we'll be investigating in this chapter and the next.

In order to describe quantum systems with high energies, and their complex interactions, relativistic quantum theory makes explicit use of the idea of a 'field,' introduced in Chapter 3. Many quantum systems can roughly be thought of as excitations of such fields.[2] Recall that a quantum field is an entity capable of being excited into higher vibrational states, like a drum head. A louder sound of the drum head is analogous to having more quanta present in the excited quantum field. The field whose excitations are the quanta of light — photons — is the quantum electromagnetic field, and the way that field gets into a higher vibrational state is by being excited by a *source* of that field. As we shall see, a source of the quantum electromagnetic field is none other than a charged quantum, such as an

[1]Of course, it's also possible to fail to describe measurement in the relativistic theory, if one is not using the transactional picture. But it's harder to ignore absorption in the relativistic theory, since emission and absorption have more equal roles in that description. For example, quantum field theory makes use of creation and annihilation operators, which are equally important in that theory. Creation corresponds to emission, and annihilation corresponds to absorption. There is nothing in the standard nonrelativistic theory that corresponds to absorption.

[2]Technically, the transactional interpretation uses a 'direct action' picture of fields, and in this picture quantum systems (as described by quantum states) can be thought of as field excitations, but the field itself is not considered an independently-existing entity. See Kastner (2012, Chapter 6) for details.

electron or proton. In this chapter, we'll study field excitations that are not fully-fledged quanta, but only 'virtual.'

When trying to picture these elusive entities (fields and their quanta), it's important to keep in mind that they are not contained in what we think of as spacetime (as discussed in Chapters 2 and 3). Rather, they exist in a realm of possibilities outside spacetime. In the next section we explore forces that arise from the relativistic domain of high speeds and energies. Forces arise from an even more tenuous form of possibility than the offer waves discussed in Chapter 3, so let's start by recalling just how physically important possibility can be.

Possibility: The Strongest Thing in the World

In going about our everyday lives, we routinely interact with material objects. We take for granted that those objects are constructed out of sturdy stuff (atoms). However, classical physics has no explanation for this stability and solidity of matter. We must look to the quantum description instead. And in doing so, the first thing we must do is rid ourselves of the notion of electrons as charged little balls. As noted in Chapter 1, this was the picture assumed by classical physics. The atom was initially thought to be a composite object that had a positively charged nucleus, and one or more negatively charged electrons 'circling' around it. A circling, or orbiting, motion is a kind of accelerated motion. According to classical electromagnetic theory, accelerating charged objects radiate energy, so those circling electrons should be radiating away their energy as they move around the nucleus. This would be a highly unstable situation: such electrons would radiate away their energy, suffer orbital decay, and crash into the nucleus in a very short amount of time. Classical physics, therefore, could not explain the basic stability of the atom, which was supposed to be the fundamental building block of matter.

Quantum physics was constructed to account for the fact that electrons are indeed bound to an atomic nucleus, and that they seem to 'go around' the nucleus in some way, yet they do not lose any energy while doing so. So an electron cannot be viewed as an ordinary charged object that behaves in all respects described by classical electromagnetic theory. It clearly has the ability to enter into a relationship with an object of opposite charge (the nucleus) in a way that allows it to occupy certain stable states, even though these states of motion would be highly unstable according to the classical theory. This relationship between two or more quantum systems is called a bound state. (We'll consider bound states in a little more detail in the next chapter.)

What is it about a quantum bound state that allows the electron to escape the fate (i.e., orbital decay) of a macroscopic (classical) charged object? The answer, in the interpretation proposed here, is that an electron is an offer wave: a physical possibility rather than a physical actuality. The electron offer wave enters into a bound state with the nucleus due to the attractive electromagnetic force between the two, which is not a transaction. The force that causes it to enter into that bound state with the nucleus is called the Coulomb force, after its discoverer.[3]

In contrast to electrons detected in an actualized transaction, the electrons bound to an atomic nucleus are not actualized transactions. These electron possibilities are still offer waves, and therefore cannot be pinned down to a particular place at a particular time. Remember, offer waves are not spacetime objects with determinate positions; they are more tenuous than that. Ironically, this very tenuousness is what allows the atomic electron to be 'spread out' around the nucleus, in a kind of cloud, rather than occupying a definite classical orbit that would be subject to the decay, by radiation of energy, described above. What is definite about the electron offer is its energy, which remains constant even though this is only possible energy. It is possible energy because there is no actualized transaction taking place. Only through an actualized transaction can real energy be radiated, i.e., transferred from one object to another.

The picture of an electron as a possibility cloud surrounding a nucleus is what gives matter its stability. This powerful efficacy of 'mere possibility' may seem strangely paradoxical. However, as discussed

---

[3] Charles Augustin de Coulomb (1706–1806) discovered the law of electrostatic attraction and repulsion between charged objects. This force is conveyed by an even more tenuous version of physical possibility than offer waves. The possibilities conveying interparticle forces such as the Coulomb force are called virtual quanta, and these are what we'll be considering in this chapter.

above, if the electron were an actual, concrete particle orbiting the nucleus, it would certainly crash into it, and that would be the end of the atom.

Now that we understand how the nature of the electron as diffuse possibility gives matter its basic stability, we can move to the solidity of atoms and matter. There are two important quantum principles that help explain the solidity or incompressibility of atoms. The first is Heisenberg's Uncertainty Principle (HUP), which you may recall from Chapter 2. HUP tells us that if we have maximum determinacy of momentum, we have minimum determinacy of position. Maximum determinacy of momentum means the quantum is in a state of definite momentum, ***p***. Having a definite momentum means a quantum can be described by a single momentum possibility triangle, say | **5** > (five units of some arbitrary measure of possible momentum), rather than a superposition of different momenta, such as | **4** > + | **5** > + | **6** > (using the notation we introduced in Chapter 2). In the case of an atomic electron, both its momentum and energy are very well defined, so its position — an observable which is incompatible with momentum — is undefined, although its distance from the nucleus does have a most likely, or average, value. (Recall also from Chapter 2 that position and momentum are incompatible observables.)

A specific example of an electron having a definite (possible) energy is an electron in the lowest energy state in a hydrogen atom. In this state, the electron is not really 'going around' the nucleus at all (Figure 4.1).[4] Such an electron is just a diffuse 'cloud' of possibility centered on the nucleus. Now, suppose you tried to squeeze that electron cloud down to a smaller cloud. This would decrease the uncertainty of its position by confining it to a smaller region. The HUP then dictates that its momentum — in this case, momentum straight inward or outward from the nucleus — would become more uncertain. With increasing uncertainty in momentum comes a greater likelihood of a larger momentum, which corresponds to more

[4] In the ground state, the electron's orbital angular momentum, the observable that measures how fast something is going around in a circular motion, is zero. This is a characteristic of all atomic ground states. Higher energy states, called excited states, can have a nonzero angular momentum. But even these are not literally orbiting the nucleus like a planet around the Sun, so again there is no orbital decay

Figure 4.1. An electron possibility cloud.

energy of motion. More energy of motion means a greater resistive force pushing outward. Therefore, the tighter a space you try to cram the possibility cloud into, the more energetically it will resist.

In addition to the HUP, electrons also obey something called the Pauli Exclusion Principle (PEP). This principle, named after its discoverer, Wolfgang Pauli, tells us that no two electrons in the same atom may have the same quantum state. In transactional terms, this means that their offer waves must have different energies and momenta. Any atom that has two or more electrons must conform to this principle. If you try to compress the electron cloud of this atom, it has the effect of pushing the electron offer waves towards the same energies and momenta. They are not going to accept that, because they are more powerful than you. The only way this quantum resistive power can be overcome is by an enormous astronomical object, such as a dying star significantly larger than our Sun, which is collapsing in on itself.[5]

---

[5] Such an astronomical collapse results in all those atoms being compressed by their collective gravitational pull into a sphere only about 15 miles in diameter, an immensely dense object. Under this degree of compression, the only way the electron offer waves can avoid occupying the same atomic state is to give up their status as independent electronic offer waves and to merge with the protons in the atomic nucleus. The process of an electron merging with a proton in this way creates an entirely new particle, the neutron. When this occurs, all the atoms in the star are converted into a big mass of neutrons. Such an object is called a 'neutron star.' The neutrons also obey the PEP, and this 'neutron gas' produces its own outward pressure that resists further compression. However, that can theoretically also be overcome by further collapse if the dying star has more than three solar masses, theoretically resulting in a black hole. (However, the existence of black holes is now controversial.)

These two quantum forces, the HUP and the PEP, apply at the level of quantum possibility rather than spacetime actuality. The HUP makes atoms incompressible by providing a resistive force against anything that would tend to reduce the position uncertainty of the electron possibility cloud. The PEP prohibits the possibility offers of two electrons in an atom from having the same energy and momentum. Yet, even though HUP and PEP act only at the level of possibility, these quantum laws work together to give the atom its stability and its resistance to being crushed. The importance of this fact cannot be overstressed: it is possibility that gives atoms their strength and stability. *The possibility clouds of electrons are the impenetrable scaffolding upon which all apparently solid matter is constructed.* Without these physical possibilities, matter as we know it would not exist, since all the electrons comprising the atoms would have long since crashed into their respective nuclei.

Forces as Possibility

In addition to the quantum mechanical principles discussed above, forces play a crucial part in the apparent rigidity of ordinary macroscopic matter. For example, as I type on the keys of my laptop, specific events, such as the writing of this book letter by letter, occur because of the forces exerted by my fingers against the keys. The atoms making up my fingers and the keys are parts of larger structures (for example, the cells in my fingers), but the forces holding the atoms and molecules together in these structures are electromagnetic. So, in addition to the quantum forces responsible for the basic incompressibility of atoms discussed above (i.e., HUP and PEP), the forces that prevent my finger from just passing through the keyboard are primarily due to electromagnetism. Classical physics can provide general laws characterizing the action of these forces, but in more fundamental terms, these forces are conveyed by possibility, and an even more tenuous form of possibility than that of the electron offer wave 'cloud' surrounding a nucleus. The more tenuous type of possibility is called a virtual quantum. In contrast, we can think of an offer wave as a persistent quantum, which is fundamentally more stable and long-lived than a virtual quantum.

The virtual quanta that convey forces become apparent only in the relativistic theory, which is capable of describing the finer details of particle interactions. The relativistic domain involves higher energies than the non-relativistic domain, and one of the most important aspects of this domain is that quanta can be created and destroyed. In contrast, the non-relativistic domain only addresses persistent quanta that are neither created nor destroyed. An electron in a stable atomic state can be well described this way. But, clearly, the latter is an idealization, since all quanta are in fact created and destroyed. The nonrelativistic theory deals only with quanta that are relatively stable and long-lived.

In the interpretation presented here, the persistent quanta addressed by the nonrelativistic theory are none other than the possibility offers with specific qualities (for example, an electron offer of momentum $\boldsymbol{p}$, which would be described in symbols as the labeled triangle $|\mathbf{p}>$). But to describe forces (like the electromagnetic force that holds the molecules of my fingers together), we need the more tenuous kind of physical possibility, the virtual quantum. This is where the particle creation and destruction of the relativistic realm enter the picture. These virtual quanta are so short-lived and uncertain that they don't rise to the level of an offer; that is, they are not the kind of entity described by a stable, long-lived state such as $|\mathbf{p}>$. However, they can make two or more offer waves interact; in a sense, two or more offer waves, such as electron offers, can toss these virtual quanta back and forth. We will see later that by tossing these virtual quanta back and forth, the electron offers actually experience a force between them. Not only electrons experience forces by exchanging virtual quanta, but all quanta.

In order to understand the relationship between virtual quanta and the more persistent offer waves, think of a coin flip. First, in the nonrelativistic realm, our coin flip has two possible outcomes: (1) heads, meaning there is an offer wave; or (2) tails, there isn't. There are no other options. This is how the nonrelativistic theory works: either we have a possibility triangle, or we don't. Now let's switch to the relativistic domain. Suddenly things become more complicated; it is as if the coin gets thicker. Now, we might get heads, or we might get tails, or a new, intermediate kind of process may happen. Figuratively speaking, the coin might land on its side (Figure 4.2). The case of the coin landing on its side is a new, relativistic

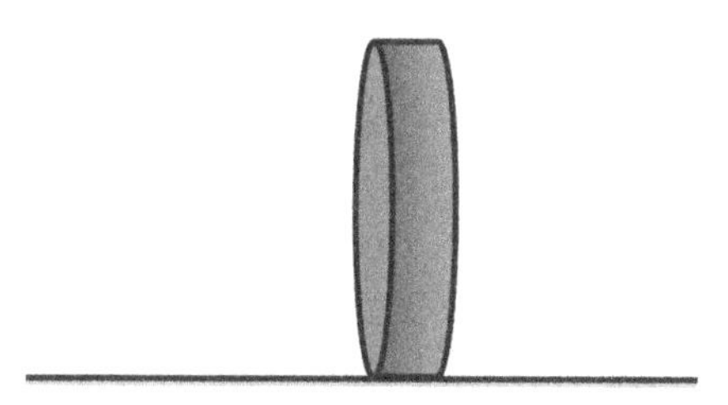

Figure 4.2. Virtual quantum: like a coin toss that results in neither heads nor tails, but something in between.

process that is like a hint of an offer wave, but which does not rise to that level. This is the nature of a virtual quantum. For example, an electron offer wave can (in principle) emit a virtual photon. The virtual photon is a quantum entity that has a latent possibility of becoming an offer, but it does not rise to that level, and therefore remains only a 'virtual' quantum. (There are other types of quanta besides electrons and photons, but for simplicity we'll focus on these.) A photon offer has not been emitted from the electron offer, but there is still something there, something in between; the nuance of an offer, if you will.

These virtual photons can have very strange characteristics. To see what they are, we need to recall the famous discovery by Einstein that $E = mc^2$. This equation, from the classical theory of relativity, describes a strict relationship between the energy, E, of an object and its mass (i.e., its amount of matter), m: it says that the energy carried by an object is equal to its mass times the square of the speed of light, denoted by c. Now, photons are massless, but they do have momentum, p, and the equivalent version of Einstein's prescription for photons turns out to be $E = pc$. In other words, this says that the energy of a photon is equal to its momentum times the speed of light. This relationship between energy and mass or momentum is known as the *mass shell condition*. Quanta that obey this relationship are said to be 'on the mass shell.' However, this relationship only applies to real quanta (either offers or quanta transferred in actualized transactions), not virtual quanta.

The bizarre thing about virtual quanta is that they can have energies that do not obey the mass shell condition, so they are neither fully-fledged possibility offers nor real packets of energy. This makes them very strange creatures indeed. Even though they appear only in relativistic quantum theory, they seem to violate a key feature of classical relativity and get away with it![6] The non-adherence of virtual quanta to this important condition, a fundamental principle of Einstein's theory of relativity, provides a further indication of how insubstantial they are. Yet, as we'll see, they give rise to very real physical effects.

We've noted above that virtual quanta are very elusive and insubstantial, but they are nevertheless immensely consequential. In this regard, an interesting and well-established aspect of standard relativistic quantum mechanics is that this tendency of a quantum system to emit (or absorb) virtual photons is an expression of an object's charge. That is to say, if an object possesses the attribute we call charge, it has this ability to emit or absorb virtual photons. This applies to any charged particles, such as negatively-charged electrons or positively-charged protons. (There are others that we will not list here.) For example, a virtual photon exchange between two charged particles is the quantum mechanical foundation of the macroscopic, classical phenomena of electromagnetic attraction and repulsion between electrically-charged objects. In other words, it is the exchange of virtual photons that causes electromagnetic attraction or repulsion.

We'll discuss the relation of virtual particle exchange to forces in more detail later on, but first let's take a closer look at the basic process. As noted above, a charged quantum system, such as an electron, has a tendency to emit or absorb photons (quanta of the electromagnetic field). A charged quantum system can be thought of as always being surrounded by a 'cloud' of virtual photons popping in and out of existence; its entourage, if you will. This 'cloud' of virtual photons — emissaries of the electromagnetic field — is an expression of the fact that the electron, as a charged object, is a source of that field. Again, these virtual photons (indicated by the curly arrows in Figure 4.3) are not offer waves. They correspond to the coin 'landing on its side' in the metaphor above. When two electrons approach each other, a virtual photon may stray from one and

[6]This is usually explained in terms of the HUP, but there is another, arguably better, way of explaining how virtual particles get away with straying from the mass-shell condition in TI. The technical details are in Kastner (2014b).

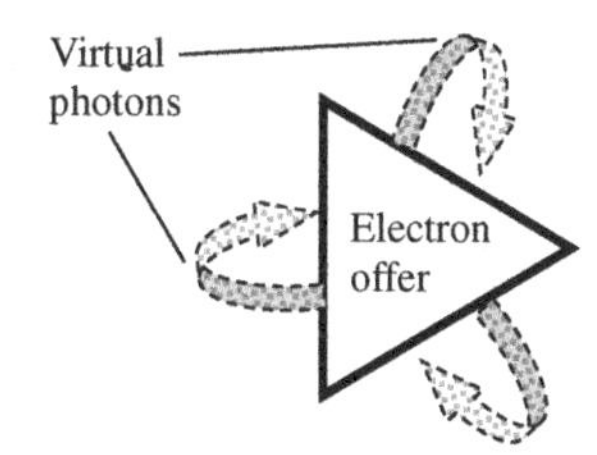

Figure 4.3. Charged quanta are surrounded by evanescent virtual quanta. The un-actualized electron is an offer wave, indicated by the possibility triangle; the virtual quanta are indicated by the arrows.

take up residence around the other; figuratively speaking, a member of one electron's entourage decides to leave and join the other. This is an example of a virtual particle exchange.

The surprising feature of this type of interaction is that there is no fact of the matter as to which electron emitted the virtual photon and which one absorbed it. That is, in a fundamental sense, neither electron clearly emitted or absorbed the photon, because no offer wave has been emitted or absorbed. All that can be said is that a virtual photon was exchanged. However, the exchange has consequences: in general, it creates a tendency for the interacting charged quanta (electrons, protons, or combinations thereof) to be either closer together (attracted) or farther apart (repulsed). If you do the relevant computations, you find that two particles with 'like' charges (i.e., both positive or both negative) repel each other due to this exchange, while two particles with opposite charges attract each other due to the exchange. Your hair stands on end due to static electricity because the like charges in your hair are repelling each other. So the next time your hair stands on end on a dry day, you can blame virtual photon exchange for it.[7]

We just noted that these hardly-there, virtual 'nuances,' having all kinds of strange energies that don't satisfy the 'mass shell' condition, create the tendencies for charged particles to attract or repel each other. Yet, despite their evanescence, these exchanges constitute the quantum-mechanical basis of all classical laws describing forces. Without them, we would not

[7] Phenomena like this, attributed to 'static electricity,' arise due to electromagnetism, which is just the name for the more general theory that includes charges in motion and magnetic fields, as well as static charges.

have the electromagnetic interactions with which we are all familiar, such as magnets sticking to our refrigerators.

Virtual Particle Exchange and Forces

We will now see how virtual particle exchanges give rise to measurable forces. We said above that forces are established through the exchange of virtual particles, more accurately described as virtual quanta. But this 'exchange' is not a transaction. What is the difference between a transaction and a virtual quantum exchange? To answer this, we will be extending the original transactional interpretation (TI) into the relativistic domain. While Prof. Cramer introduced the basic TI in the 1980s (1983, 1986), the material we'll be discussing below was introduced by this author in 2012.[8]

As noted above, in the case of a virtual quantum exchange, there is no fact of the matter about which quantum system emitted the virtual quantum and which one absorbed it. So there is no confirming response from either of the quantum systems exchanging the virtual quantum. A confirming response can only happen when there is a fact of the matter about which quantum system is the emitter and which the absorber, with the emitter clearly generating an offer and an absorber generating a confirmation in response to the emitter's offer. However, in a virtual quantum exchange, neither emission nor absorption of an offer has really occurred (the coin has landed on its side); so the basic quantum uncertainty remains, and there is no specific event actualized.

Another way to understand this process is in terms of possible energy: in a virtual quantum exchange, there is no transfer of real energy from one quantum to another; the energy remains at the level of possibility. Yet again, this 'possible' energy has real physical effects in that it creates a measurable tendency for two such interacting quanta to be either attracted or repulsed. I refer to these effects as 'measurable' because they underlie all the everyday, macroscopic electromagnetic phenomena with which we are all familiar, and which can be well predicted by classical electromagnetism.

---

[8]Readers interested in the technical details of this new proposal may wish to consult Kastner (2012, Chapter 6, and 2014a,b).

The way these virtual photon exchanges create measurable effects is by changing the offer waves corresponding to the interacting quanta (for example, electrons). In doing so, they increase the probability of transactions corresponding to behaviors of attraction or repulsion (for opposite or like charges, respectively). This is the new way of looking at TI in the relativistic realm. Prior to this, TI (as originally proposed by Prof. Cramer) explained the basic process of measurement in terms of offer waves (OW) and confirmation waves (CW), but it did not explain how those OW and CW were affected by virtual quanta, or how the OW and CW are produced in the first place (the main topic to be explored in the next chapter).

To make clear how the virtual quanta influence the probabilities of available transactions and thereby underlie the classical laws, let's consider a specific example. We could set up two different electron sources that could emit two electron offers in close proximity. Remember that electrons, being of like charge, repel each other. These two electron sources could be aimed so that electrons would be detected some distance away at a screen comprised of detectors (small but macroscopic absorbers). This scenario is illustrated in Figure 4.4. Figure 4.5 shows, in schematic terms, that the exchange of a virtual quantum between two electron offer waves introduces

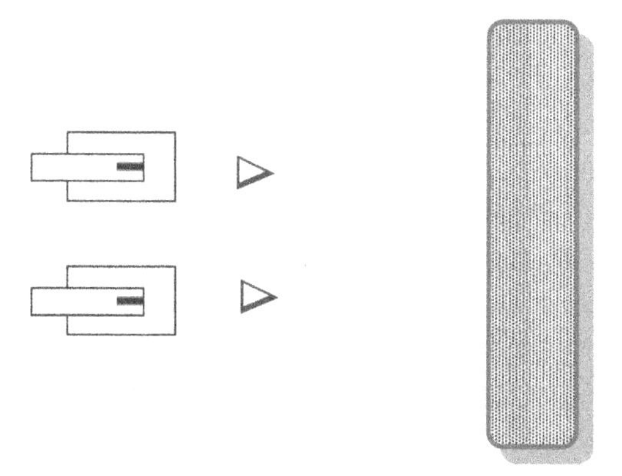

Figure 4.4. Two electron guns send out offer waves with initial momenta heading straight toward a detection screen. The small dots on the screen indicate small but macroscopic-sized absorbers that can indicate the position on the screen at which an electron is absorbed.

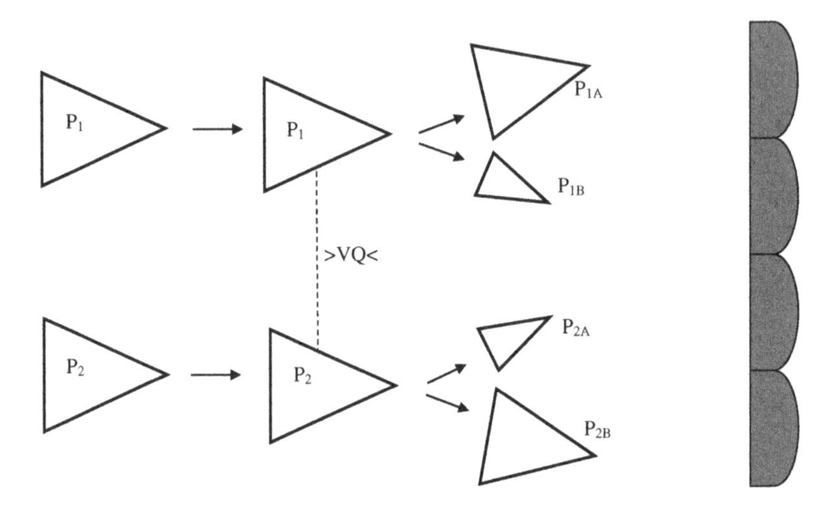

Figure 4.5. Two electron offer waves exchange a virtual photon (>VQ<). This results in a change in the offer waves. They develop amplitudes for momenta other than their original one. The shapes on the right represent close-up views of individual macroscopic absorbers in the screen.

a new kind of uncertainty in both interacting offer waves. It creates new amplitudes for different offers to reach the final absorbers; that is, different offers from the ones that were emitted from the source.

As depicted in Figure 4.5, two electron offers start out from their sources with specific momenta, P_1 and P_2, and then they exchange a virtual photon. (Actually, electrons are continually exchanging virtual photons, but we're just considering a single exchange for simplicity.) This results in a new uncertainty, in which each of the original electron offers transforms into a superposition of offer wave components with varying amplitudes. Thus, P_1 becomes a superposition of P_{1A} and P_{1B}, and likewise for P_2. (In general, there are many other components; we just show two here for simplicity.) After such an exchange, the inital offers develop larger amplitudes for momentum states directed away from each other.[9] So here we see that the largest amplitudes are for the new offer waves P_{1A} and P_{2B}, which are pointing away from each other. This is the quantum-mechanical origin of all forces.

[9]Actually, the repulsion is established by the phases, rather than the magnitudes, of the amplitudes of the scattered offers, but those details are beyond the scope of this book.

As discussed in Chapter 3, a competing set of incipient transactions is then set up between each electron source and the different absorbers in the screen, each with different probabilities of actualization. If we let our two sources emit electron pairs like this over and over, we would find that the most probable transactions — i.e., the ones occurring most frequently — would correspond to electron paths moving apart by the amount corresponding to their electromagnetic repulsion, which was brought about by the virtual particle exchanges discussed above. If there were no such virtual quantum exchanges, the most probable transactions would be the ones corresponding to detections straight in front of the electron sources, instead of the ones corresponding to paths moving apart. Forces act on quantum objects (possibility offers) through even more subtle quantum objects (virtual quanta), by increasing or decreasing the probabilities of the various transactional opportunities available. This is the basis of the classical phenomenon of electrostatic repulsion of the electrons coming from the two sources. By extension, it is also the basis of the nuclear forces that hold atoms together and allow matter to exist. In other words, matter could not exist without the nuclear and electromagnetic forces.

Another way to visualize this basic attraction or repulsion process is by using a diagram invented by Richard Feynman. Since Feynman contributed so much to relativistic quantum theory, including these very helpful diagrams, they are named after him.[10] In a Feynman diagram, the interaction in which a virtual quantum comes into being can be visualized as a vertex where three lines come together. In Figure 4.6, the upward arrows represent an offer wave, such as an electron, that can give rise to virtual photons. (The arrows are bent to illustrate that the electron offer has been altered in some way; not necessarily repulsed, like in the preceding example.) This propensity of an offer to be altered through an interaction involving virtual quanta is called *coupling*, and it is described numerically

[10]Readers familiar with Feynman diagrams in the context of standard quantum theory may wonder about their use in the transactional picture. Using them here is perfectly legitimate, since TI uses all the standard entities of quantum theory, such as quantum states, in the same basic way. In TI, the usual quantum state is an important part of the story, but not the whole story. Readers interested in the technical side of the use of Feynman diagrams in this picture may refer to the work of Paul Davies, who extended the Wheeler–Feynman picture into the quantum relativistic domain (Davies, 1971, 1972).

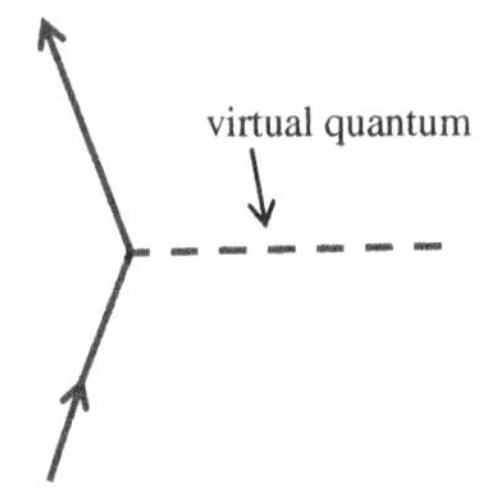

Figure 4.6. An offer wave (arrow) gives rise to a virtual quantum (dashed line).

by a number called the *coupling constant*. The coupling constant turns out to be the amplitude for an offer to emit or absorb another kind of offer. The virtual quantum is designated by a dashed line.

This could represent an electron offer wave emitting a virtual photon, but there are other kinds of quanta that can emit virtual particles as well. To clarify, to 'emit a virtual photon' means that the coin has 'landed on its side': a photon offer wave has not been emitted, but there is still a sort of 'wisp' of the electromagnetic field there. That fleeting wisp is the virtual photon. However, it's important to note that such an interaction never occurs in isolation. It is always accompanied by another interaction, also characterized by the coupling constant, which is a possible absorption of the offer. (Remember that 'coupling' is the tendency to emit or absorb.) Figure 4.7 is a sketch of how these two 'mirror' processes always occur together to create one exchange of a virtual quantum, called a 'tree' process.

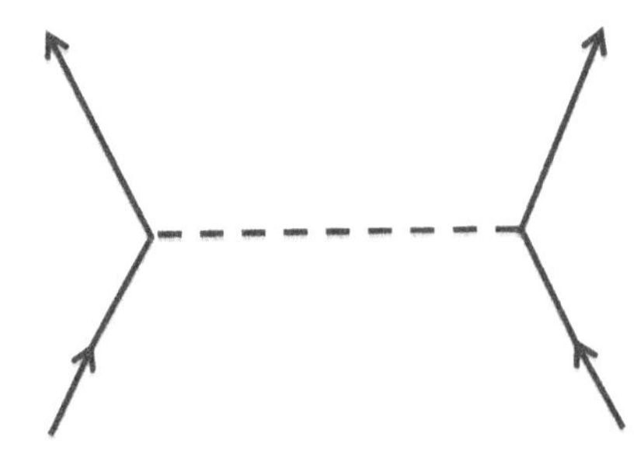

Figure 4.7 Two quanta exchanging a virtual quantum. The changing directions of the arrows indicates that the offer wave (quantum state) of each may change due to the interaction. The dashed line is a virtual quantum. There is no arrow on the virtual quantum because there is no fact of the matter concerning which of the offers emitted it and which one absorbed it.

As noted previously, at this relativistic level there is no fact of the matter about which one of these coupling processes is the 'emission' and which one is the 'absorption.' This is because there isn't really a definite emission or absorption in this process (that's the defining nature of the exchange of a virtual quantum). That is, it is just as likely that the quantum on the right is the possible emitter and the one on the left the possible absorber, or vice versa, since each of these is only a possible emission or absorption. So there is no confirming response, and no transaction occurs between the two quanta. But again, something does happen, albeit on a very subtle and sub-empirical level: a transfer of possible energy. This is indicated by the fact that the participating quanta 'scatter'; that is, they affect each other at the level of possibility. In other words, there was a force between them.

We should also note here that there is also a Feynman diagram representing an electron offer wave surrounded by its virtual photon 'entourage,' as mentioned earlier. In this case, one or more virtual photons can 'pop' in and out of existence while coming and going from the same electron offer; this is called self-action (Figure 4.8). The tree diagram and the self-action diagram represent the exchange of a virtual quantum, in which there is no confirmation and thus no transaction. (The single electron depicted in the self-action diagram above is considered as exchanging virtual photons with itself.)

Thus, we see that virtual particle exchanges are crucial for conveying the influences of forces, but these exchanges are not transactions; the virtual particles do not rise to the level of offer waves or generate the confirming responses of absorbers. They can also have strange energies that do not satisfy the 'mass shell' condition, and this is another reason why they do not rise to the level of offer waves or generate confirmations.

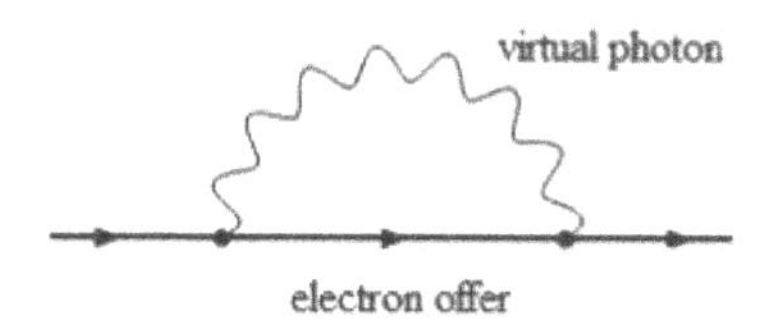

Figure 4.8. An electron 'exchanges' a virtual photon (indicated here by a wavy line) with itself.

As noted above, offer waves (and therefore their responding confirmations) must satisfy the mass shell condition, $E = mc^2$ (or $E = pc$ for photons).

In this chapter, we saw how electrons, and other subatomic particles, are not little charged particles, but are instead offer waves. These offer waves give rise to transactions, and it is the actualized transactions that cause energy transfer. We saw how the HUP and PEP act at the level of possibility to give matter its solidity and form. We encountered virtual quanta and saw that they are subtle tendencies that appear only in relativistic quantum theory, and how they convey the electromagnetic and nuclear forces by changing the offer waves. Virtual quanta do not satisfy the 'mass shell' condition and therefore cannot transfer real energy, but they play a crucial role in that they give rise to forces that influence the probabilities of various transactions, and this in turn influences the observable behavior of objects and events actualized through those transactions.

In the next chapter, we'll consider the following question: What conditions are necessary in order to generate offers and their confirmations from the ephemeral relativistic foundation from which possibility is born?

Chapter 5

From Virtual to Possible to Real

> *'Between the idea*
> *And the reality*
> *Between the motion*
> *And the act*
> *Falls the Shadow'*
>
> *T. S. Eliot, The Hollow Men*

In the previous chapter, we saw how forces are conveyed by transient virtual quanta that are exchanged between offer waves. For example, virtual photon exchanges are going on all the time between the offer waves of charged quanta such as electrons and protons. These virtual quanta, which arise from quantum fields, are even more tenuous forms of possibility than the more persistent offer and confirmation waves. Yet they influence the behavior of those offers in a way that results in the measurable phenomena described by the laws of various forces.

In this chapter, we continue in the relativistic realm, and discuss more aspects of the extension of the transactional interpretation (TI) proposed by this author. This new extension consists of showing how the offers and confirmations also have their origins in the underlying, incessant activity of the quantum fields that are the sources of virtual quanta.

A Brief Review

We first need to review the possibility triangles, or quantum states, introduced in Chapter 2. These represent offer waves, and they are not limited to the nonrelativistic picture. They are also part of relativistic quantum mechanics, and in that context they describe fully-fledged excitations of their underlying quantum field, as opposed to just the virtual quanta described in the previous chapter.[1] Recall that these excitations are what you get when you jump on those quantum field 'drum heads' and get them excited into persistent vibrational states. These persistent vibrational entities are the offer waves that can toss virtual quanta back and forth. This process was discussed in the previous chapter, with the example of the two electron offer waves that were modified by their virtual photon exchanges (illustrated in Figure 4.5).

In a virtual particle exchange, a force is exchanged between two offers, but no energy is transferred. As we've seen in the previous chapter, the force acts by changing the possibility offers that are participating in the virtual quantum transfer, so that their transactional opportunities are affected. This in turn affects the behavior of the offers, which is what forces do; and they don't actually need to transfer energy in order to do it! Note also that the influence of the force is symmetrical: the virtual quantum transfer affects both of the interacting quanta in the same way. It doesn't matter which of the quanta you think of as 'acting on' the other: they are acting on each other via the transferred virtual quanta they are exchanging. And again, there is no fact of the matter about which quantum is emitting any one of the virtual quanta and which is receiving it.

However, in a transaction, real energy is transferred from one quantum to another in an asymmetrical process. In this case, one quantum is really emitting an offer wave, while one or more other absorbing quanta are responding with confirmations. This sets up a competing set of incipient transactions, as described in Chapter 3. Upon actualization of one of the incipient transactions, the emitter loses some amount of energy, and the 'winning' absorber, which is called the *receiving* absorber, gains it. In this chapter, we'll see how the symmetrical process of virtual particle exchange described in Chapter 4 can be transformed into a very different and asymmetrical one. That asymmetrical process is the emission of an offer wave, one or more confirmations, and an actualized transaction resulting in energy transfer from the emitter to the receiving absorber.

---

[1]These fully-fledged excitations are traditionally called 'real quanta.'

How do Transactions Occur?

Two electron offer waves may interact through the exchange of virtual photons, and we have seen that this kind of interaction is the basis of forces. Yet, as noted above, these virtual particle exchanges are not offers, confirmations, or transactions. Sometimes, however, instead of a virtual quantum, an offer wave can be spontaneously emitted, which leads to one or more confirmations. This in turn sets up one or more incipient transactions, which results in an actualized transaction and real energy transfer. In order to have a fully-fledged offer and confirmation, as opposed to merely the transfer of virtual quanta, the objects involved have two big hurdles to overcome. One hurdle concerns the conservation laws for energy and momentum, which must be fulfilled in any transaction. The other hurdle concerns the nature of coupling, which is the ability of one kind of quantum to emit or absorb another kind of quantum. We examine these hurdles in detail in what follows.

The first hurdle: Energy conservation

First, what exactly is energy conservation? It is nature's requirement that the energy 'books' always have to be balanced.[2] Much like money, in order for energy to be transferred there has to be a supply, and there has to be a place to put the amount transferred; that amount cannot simply disappear. (In fact, this kind of financial transaction, in which the recipient plays a crucial role, is the process that gave the TI its name. To further the analogy, the quantum offer and confirmation can be seen as a negotiation that precedes the transaction.)

The energy supply is provided by a quantum system that has excess energy that it can give away and still end up in a stable state after doing so. This is analogous to the way in which you can safely pay a bill only if you have enough money sitting in your bank account. In order for that

---

[2]In technical terms, energy conservation is the law requiring that energy can neither be created nor destroyed. However, this strictly refers to 'real' energy transferrable in spacetime. Thus it does not apply to virtual quanta, which do not exist in spacetime and cannot transfer energy.

energy to be transferred, there also must be another quantum system that can gain that exact amount of energy and end up in a stable configuration after doing so. As you may have guessed, the first is an emitter and the second is an absorber. Typically, emitters and absorbers are electrons in atoms, which can emit and absorb this energy in the form of photons, quanta of the electromagnetic field.

As an example, consider the emission of a photon from an electron in an atom. Before the electron can give up the energy, it has to be in what's called an excited state, which is any state with more energy than the lowest possible energy state (which is the called the ground state). Again, this is because it has to have some excess energy to give up and a lower, stable energy state available for it to retire to after emitting the photon. If the electron is already in its ground state, it cannot emit a photon and thereby lose energy, since there is no lower energy state available to it. Similarly, the absorbing electron must have a stable higher energy state to land in upon absorbing the energy. That energy must be used by the entity that absorbed it to create some stable, new configuration. Each possible state of the electron is characterized by a certain specific amount of energy. So, for example, the excited state could be characterized by the energy E_1, while the lower energy state could be characterized by E_2.

In addition, the energy transferred by the photon must be just the right amount to take each of the participating electrons from their respective initial to final states. That energy must be equal to the difference between the energies of initial and final states of the electrons that emit and absorb the photon. So, for example, if the initial and final energies of the emitting electron are E_1 and E_2, then the photon must have energy $E_1 - E_2$.[3]

The second hurdle: Coupling

The second hurdle is related to coupling. It is important to note that the term 'coupling' applies to both virtual and real quanta. The coupling constant is the quantity that describes the strength of the force involved in the interaction of charged objects via virtual quanta (e.g., two electrons

---

[3]A related requirement is that the emitted photon must be on the mass shell (recall Chapter 4). The mass shell condition says that a photon with energy $E_1–E_2$ must have momentum $(p_1–p_2)c$.

repelling each other).[4] But now we need to consider another important physical meaning of the coupling constant. It turns out that the coupling constant is a special kind of amplitude that comes into play only at the relativistic level.

We previously discussed amplitudes in connection with the sizes of the possibility triangles of Chapters 2 and 3. Recall that an amplitude is the quantum analog of classical probability. The Born Rule tells us to square the amplitudes of the possibility triangles (more precisely, take their absolute square) to find the probability that the property labeled by that possibility triangle will be the one that is observed. In the transactional picture, the Born Rule arises because there is an amplitude associated with both the offer and the confirmation. Recall from Chapter 3 that the offer and confirmation amplitudes are 'mirror images' of each other (complex conjugates). When these are multiplied together, we get the absolute square of the amplitude of the offer wave, and this gives the probability that that particular transaction will be actualized. That is how the transactional picture provides a physical reason for the squaring procedure of the Born Rule.

The coupling constant is another kind of amplitude that operates only at the relativistic level: in traditional terms, it is the amplitude for a charged particle to emit or absorb a real photon.[5] Recall that virtual photons are being exchanged all the time between charged particles such as electrons. When the conditions are right, what would have been just a virtual photon is elevated to a 'real photon.' In the transactional picture, a 'real photon' is an offer wave whose absorption generates one or more confirmation waves, depending on the absorber configuration. Therefore, the coupling constant — now we'll use the more descriptive term 'coupling amplitude' — is the amplitude for any microscopic quantum to either emit an offer wave or respond, with a confirmation, to another

[4]The term 'charge' has a more general meaning than the usual electrical charge. It can also describe couplings between other kinds of fields, for example in interactions mediated by the 'strong' and 'weak' nuclear forces.

[5]That is actually the way the standard relativistic quantum theory is interpreted. Feynman himself noted that the coupling amplitude in quantum electrodynamics is the amplitude for a real electron to emit or absorb a real (as opposed to virtual) photon (Feynman, 1985, p. 129).

quantum's offer wave. When that happens, what would have been 'just a virtual photon' is instead emitted as a fully-fledged offer wave, is responded to by confirmations, and results in a transfer of energy.

It is very uncertain (even if all the conservation laws are satisfied in the first hurdle) that any individual electron will either emit or absorb a photon offer wave, since the coupling amplitude for this process in the electromagnetic interaction is rather small: it is roughly equal to 0.085. But again, neither emission nor absorption occurs in isolation.[6] When we take this into account, the relevant probability that a photon offer will be both (1) emitted by an emitter and (2) confirmed by an absorber gives us two factors of the coupling amplitude. This means that we have to square it, giving us a basic probability of 0.007, or roughly 1/137 (less than 1%), that a photon offer wave will be emitted and absorbed. When this occurs, it means that a would-be virtual photon has instead been emitted as a real photon, which in TI is a confirmed offer wave that can be described by the nonrelativistic theory. That is, it can be represented by the 'bow tie' of Chapter 3.[7]

This possibility, the elevation of a virtual photon to an incipient transaction (with a chance of actualization and energy transfer), is not part of the standard approach to quantum theory. In fact, it is the 'missing link,' the bridge between relativistic quantum theory (which deals with virtual photons) and nonrelativistic quantum theory (which can deal only with offers and confirmations). It serves as a crucial connection point, provided only by the transactional picture, between those two theories. It's important to note, however, that in a realistic situation there will be many absorbing systems interacting with a single emitting system. Therefore we will have a set of incipient transactions corresponding to all the different directions in which the emitted photon could go. In other words, in general the photon is really a set of incipient transactions, only one of which will be

---

[6]We cannot consider the emission of an offer as an isolated process; if an offer is emitted, it must always be confirmed. This is how the transactional picture works. The offer is always only part of the story.

[7]The standard approach will often refer to a photon in a quantum state (i.e., a possibility triangle rather than a bow tie) as a 'real photon.' This is because the standard approach does not recognize confirmation. For that reason, the terminology 'real photon' is a bit ambiguous.

actualized. The quantum of electromagnetic energy will be delivered to that one 'winning' absorber, which (as noted at the beginning of this chapter) we call the receiving absorber.

At this point, the reader may be thinking: 'Hold on a minute. You have two electrons just tossing virtual photons back and forth, but then all of a sudden, one of these becomes an offer wave that can receive confirmations from possibly many electrons?' Yes, but only under the right conditions. The reason that this can happen is that electrons are always interacting via virtual photons, which are emissaries of the underlying quantum electromagnetic field. The field serves as a kind of 'sounding board' among all electrons, just as the body of a violin can resonate based on the smallest hint of a vibration of one of its strings. In the same way, the field serves as a resonator that communicates the presence of an excited electron to many ground state electrons that could receive its energy. Whether or not such a process occurs is inherently unpredictable, but it can be thought of as a kind of tendency that builds up until it burst forth, like a lightning strike.

The probability, just based on the coupling constant, of a virtual photon being elevated to a set of incipient transactions is 1/137 (it's actually much lower than this when we also take the first hurdle, energy conservation, into account). This number, 1/137, is a very well-known number in physics. It is called the 'fine structure constant.' The fine structure constant is important because it characterizes the strength of the electromagnetic interaction between charged particles. However, in the TI as presented in this book, the fine structure constant gains an interesting new physical meaning: it expresses the likelihood that a photon will advance from being merely 'virtual' to being a set of incipient transactions, one of which will be actualized, and which can then transfer energy from one charged quantum to another. While a virtual photon conveys electromagnetic force, as described in the previous chapter, only a real photon can carry electromagnetic energy.

Overcoming both hurdles

To see both hurdles in play, consider the following example of an excited-state electron, we'll call it '**E**,' and another single ground-state electron,

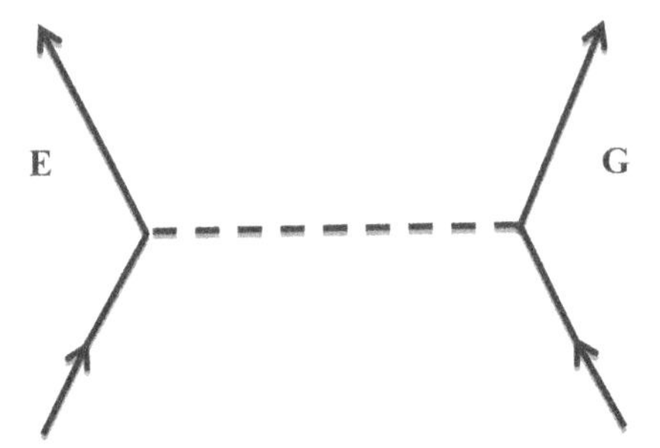

Figure 5.1. A virtual photon exchange (dashed lines) between an excited electron, E, and a ground state electron, G.

'**G**' (in a different atom). They could participate in a 'tree' type exchange as presented in Chapter 4, subject to the additional condition that they have suitable final states available (Figure 5.1). **E** is a bona fide emitter candidate because it is in an excited state and has a possible lower energy state to fall down to, and **G** is a bona fide absorber candidate, since a stable final state could result if **G** gained that amount of energy (i.e., they would exchange roles). Now, **E** and **G** are exchanging virtual photons all the time; that is, they are always interacting via the electromagnetic force. Such an exchange is pictured in Figure 5.1. However, each one of these exchanges has (at best, if all conservation laws are satisfied) no more than a 1/137 chance of resulting in a transaction.[8] So that means that **G** is not a reliable absorber. This is because a reliable absorber is something that definitely responds to an offer with a confirmation, and **G** can't be relied upon to respond with certainty.

In addition, as mentioned previously, there are other factors besides the 1/137 that go into calculating the probability that an electron will emit a real photon. The other factors concern the nature of the candidate photon

[8]As we've noted, in the more general and realistic case, there would be more than one potential absorber for **E**. Thus, if a virtual photon were to 'survive the odds' and be definitely emitted, there would be an incipient transaction corresponding to each of the absorbers. At that point there is still another roll of the quantum dice corresponding to which of the absorbers actually receives the energy carried by the photon (this is the actualization of one of the incipient transactions). In the fully realistic case, the emitting electron is surrounded by absorbers and has to overcome additional odds relating to the specifics of its possible transition from an initial state to another final state.

itself — specifically, whether it is on the 'mass shell,' as discussed in the previous chapter and in footnote 3, and, if so, if it has the right energy — and the availability of final states to which the electron could retire. These factors always decrease the probability further. So the actual probability is generally considerably less than 1/137. Thus, no individual microscopic object really qualifies as a reliable absorber, since that is an object that (as the name suggests) can be safely relied upon to respond to an offer with a confirmation. Similarly, no individual microscopic object qualifies as a reliable emitter; the same arguments apply. Here's a rough analogy: if I throw a ball, it will only come back to me with certainty if there is a really big wall available to bounce it back to me. At the quantum level, the ball could not even be reliably thrown unless there were a lot of people like me trying to do it, in addition to a really big wall to bounce it back. Similarly, no individual quantum system can be safely relied upon to either emit or absorb with certainty.

Filling in a Missing Link

The fact that individual quantum systems are not reliable emitters or absorbers fills in the missing link between the microscopic and macroscopic realms that has created a big problem for the standard (non-transactional) ways of approaching quantum theory. This notorious problem is none other than the 'Heisenberg Cut,' as discussed in Chapter 2. Objects belonging to the microscopic realm can be in a quantum superposition, while those in the macroscopic realm cannot. Recall from Chapter 2 that a quantum superposition is a situation in which an object is described by a combination of properties that seem to be mutually exclusive. For example, in the two-slit experiment, an electron offer wave can be in a superposition of 'going through slit A' and 'going through slit B'; there is no fact of the matter about which slit it went through, although 'common sense' (which we inherit from our classical, macroscopic experience) seems to tell us that it must have gone through one or the other slit.

Let us now see more closely how this missing link — i.e., the fact that you need many microscopic systems to achieve reliable emission and absorption — works to provide a criterion that distinguishes the microscopic realm from the macroscopic realm. Let's assume that we have a reliable emitter (such as a laser). The laser is reliable because it has many identically-excited atoms, so at least one of the atoms in the laser is going to emit at any given time. In order to have a situation in which we can say that at least one absorber is definitely present for an emitted offer from the laser, we need to have a large number of potentially-absorbing quanta available, like a detection screen. This way, the quanta (both emitting and absorbing) collectively increase their chances of a transaction.

To gain insight into this idea, consider a lottery. If you buy a ticket on your own, the odds against your winning are very high. But if everyone in your office buys a ticket and you all agree to share the winnings, the chances of winning are increased according to the number of people in your office. It doesn't matter which of you has the winning ticket; if anyone in your office wins, the whole office wins. In the analogy, the entire office is the reliable emitter or absorber.

If you work out the numbers, it turns out that even a very tiny macroscopic piece of absorbing material easily beats the odds we've described above, and therefore qualifies as an absorber. For example, a piece of metal of about one cubic centimeter contains about 10^{23} conduction electrons, each of which could absorb a photon. (These conduction electrons are not bound to any particular atom, they just swarm around on the surface of the metal. This is what makes metals good conductors of electricity.) Now, 10^{23} is a huge number — a one with 23 zeroes after it — and that is way more than is needed to ensure that there is at least one 'lottery winner' somewhere in that sample of metal. In fact, even with far fewer potential absorbers we are in the macroscopic realm. If you took about 100,000 atoms in their ground states, they would make up roughly the width of a human hair. This many atoms would also virtually guarantee that absorption occurs. The retina of our eye, having many receptor cells, is also a reliable absorber.

Distinguishing Between the Microscopic and Macroscopic Worlds

Let's take a moment to review these issues, since they are crucial to understanding why the TI provides an unambiguous demarcation between genuinely quantum (microscopic) objects and the ordinary macroscopic, classical world of experience. TI explains why we need to use quantum theory for certain processes — these being the genuinely microscopic ones — but also why we can apply classical physics to other processes, which are genuinely macroscopic. The reason TI can do these things is because it takes into account absorption as a real physical process. In doing so, it provides a clear account of what constitutes a measurement, and why we apply the heretofore mysterious 'Born Rule' to calculate the probabilities of outcomes of a measurement. And it turns out that the processes constituting measurement result in the 'collapse of the wave function' long before the quantum superposition can infect the macroscopic level.

We have examined above the concept of an 'absorber.' In order to understand what makes an object an absorber, we needed to consider the deeper, relativistic level of the inherently uncertain nature of quantum possibilities, which introduces an even subtler level of uncertainty. At this level, neither emission nor absorption are automatic or assured. Rather, they are both tendencies, and these tendencies can be understood as the swapping of virtual quanta (virtual photons were discussed above as a specific example). The virtual quanta convey forces, but do not participate in the energy-transferring transactions that herald the distinction between the microscopic and macroscopic realms, unless they are elevated to offer waves ('real photons'). Thus, we found that there are both potential emitters and absorbers everywhere, but not all of them can rise to the level of a functioning emitter or absorber.

In order for a potential emitter and absorber to participate in the transactional process, a basic prerequisite is that a stable final state be available for the result of such a transaction. Above, we noted that an atomic electron in an excited state is eligible to emit a photon because it has a lower energy state to 'retire' to. However, an atomic electron in a ground state is not eligible to emit a photon because it has no lower state available, so the conservation laws will not allow such a transaction to occur. On the other hand, an atomic electron in its ground state could act as an absorber: it could absorb an incoming photon offer, gain real energy from a transacted photon, and jump up to a higher energy state. However, it can only do this if the photon is of exactly the right energy for this transition, again because of the conservation laws.

Thus, whether or not a quantum object is eligible for entry into the 'emitter' or 'absorber' raffle depends on the energy states available to it. But it's still a raffle — that is, each emitter and absorber candidate still has to surmount (in general) rather high odds in order to 'win,' i.e., participate in a transaction. Therefore, large numbers of quanta are needed to ensure that emission and absorption occur. This fact is what lets TI make a non-arbitrary distinction between the microscopic and macroscopic realms. One might ask: so exactly where do you draw the line? The answer is that in TI the physical phenomena do not abruptly change at some magical, arbitrary line. Rather, there is a kind of transition zone. The more absorbers and emitters we have, the more certain transactions become, and the better the macroscopic classical laws will describe what is happening. When we have fewer emitters and absorbers, our macroscopic picture is not as accurate, and we need to use the quantum mechanical description.

When enough candidate absorbing quanta are available to an emitted offer, the object they comprise can be considered a reliable absorber, since it is virtually certain that at least one of its component quanta will respond with a confirmation, resulting in collapse. If there are fewer quanta comprising an object, then transactions involving the object are not assured. So the need to have many quanta 'working together' gives us both guaranteed collapse and a clear criterion for the distinction between the microscopic and macroscopic realms. The microscopic realm is where emission and absorption are unlikely, and the macroscopic realm is where they are virtually assured.

Bound States: Key Players in the Creation of Spacetime Events

We now need to investigate an important quantum mechanical concept: the *bound state*. The bound state is a quantum mechanical entity that is more complex than a simple offer wave, but which does not consist of enough emitters or absorbers to qualify as a macroscopic object. It is the foundation of most of the phenomena that we experience.

First, recall that offer waves can be thought of as excitations of quantum fields. For example, a photon offer wave is an excitation of the quantum electromagnetic field. Also, an electron offer is an excitation of another kind of quantum field. But there are some genuinely quantum objects, such as atoms, that are not described by excitations of a particular field; for example, there is no 'quantum field of the hydrogen atom.' Yet an atom is certainly a quantum object: it is a system composed of other quantum objects, specifically a nucleus (comprised of protons and neutrons) and one or more electrons. This sort of complex quantum object is a bound state. It is not a single offer wave, but a tightly-interacting set of offers. These component offer waves of the atom (the protons, neutrons, and electrons) are bound together by attractive forces conveyed by virtual quanta, as discussed in Chapter 4. A still more complicated kind of bound state is a molecule, which is two or more atoms bound together by the same kinds of forces.

Bound states play an important role in bringing about observable spacetime events. To understand this, we'll first need to consider how objects become localized via transactions. To be 'localized' means to be confined to a particular region of spacetime. Since an offer wave is not a spacetime object, it's not localized in spacetime. A quantum only becomes localized in spacetime due to an actualized transaction. So, for example, a laser emits photon offer waves and these offer waves generate confirmations from a detection screen, but it is only upon actualization of a transaction that an actual photon — a quantum of electromagnetic energy — is delivered from the laser to the detection screen (see Figure 5.2). In this way, the photon becomes localized. That is, it becomes confined to a well-defined region of spacetime.

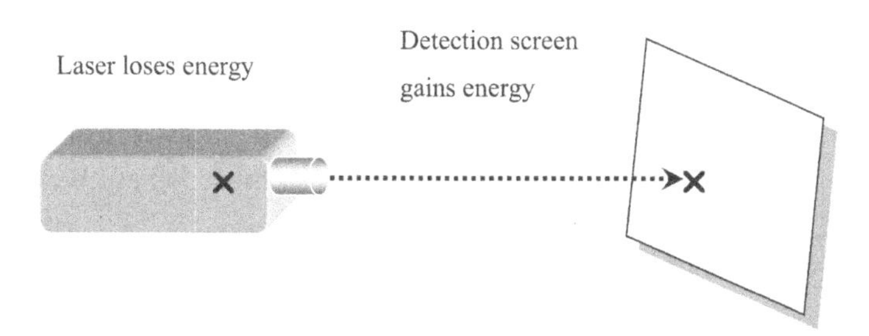

Figure 5.2. A transactional actualizing of a photon as a spacetime interval and localizing the emitter and absorber as spacetime objects. **X** indicates the localization of the emitter and absorber.

The localized photon is just this actualized transaction. It is not a little particle moving through space on a point-by-point trajectory, but it does correspond to a well-defined *spacetime interval*. The spacetime interval is a kind of separation in spacetime, defined by the emission and the absorption of the photon. (We'll discuss this further in Chapter 7.) The actualized photon is what establishes this separation, but it does not itself contain or traverse spacetime points or events. In a sense, if it 'traverses' anything, it traverses Quantumland, and that aspect of Quantumland serves as a kind of structural scaffolding to 'hold up' spacetime.

If this is hard to visualize, one can think of liquid water freezing into ice. The liquid water represents the more ephemeral entities in Quantumland and the ice represents the more rigid entities in spacetime. When water freezes into ice, water molecules become rigidly connected to each other by chemical bonds. In this analogy, two water molecules correspond to the emitter and absorber, while the actualized photon corresponds to the chemical bonds between those two molecules. The sense in which the photon does not follow a spacetime path or trajectory, as noted above, is that the chemical bonds are not made up of water molecules themselves; rather, they are the *connections* between molecules that give the rigid structure to the ice.[9]

Thus, when the photon is actualized, there is a well-defined fact of the matter as to when and where the photon originated, and when and where the photon was absorbed.[10] The **X**s in Figure 5.2 indicate this localization of the emitter and absorber in spacetime via the actualized transaction. The dotted line indicating the photon's transfer in Figure 5.2 should be considered just a connection between the emitter and the absorber that establishes their spacetime relationship, as noted above.

It's important to remember that prior to the actualized transaction, the emitting and absorbing atoms or molecular bound states were not spacetime objects; that is, they were not part of spacetime. Where were they?

---

[9]More will be said about the creation of spacetime intervals from Quantumland in Chapter 8.

[10]Technically, in view of relativity theory, each observer in a different state of motion would describe those locations differently. But they could corroborate their observations in what is called a covariant description. They would also all agree on the spacetime interval between emission and absorption. Also, in this picture 'when' and 'where' is defined relationally; that is, in terms of which entities emitted and absorbed the photons.

They were possibilities in Quantumland. This is not to say that they were not real. We have seen in this book many examples of real objects that do not exist in spacetime: offer waves, confirmation waves, and virtual particles, to name a few. So the bound-state emitter and absorber, while existing as 'only' possibilities in Quantumland, were always real, nonetheless. They were simply not spacetime objects. The actualized transaction brought these microscopic emitters and absorbers from the quantum world of possibility into the actualized spacetime world of classical physical laws.

Let us pause a moment to consider this rather astonishing new picture of reality. Everything around us that we can perceive is the result of a specific actualized event, such as the emission or absorption of a photon in an actualized transaction. All these observable events are established only through actualized transactions. This is the world of appearance, which we call 'spacetime.' But all those events are brought into spacetime from the vast hidden reality of Quantumland, which exists as well. In somewhat the same way, the Sphere did not exist in Flatland unless he was actually in contact with it. Even then, he was only intersecting it; Flatland can never contain the entirety of the Sphere. There is no time or space in Quantumland, yet it is the essential, unseen scaffolding that supports our spacetime world of experience. We'll consider some of the philosophical implications of this new picture of reality in the next chapter.

We still need to see how it is that bound states participate in transactions and become localized. To address this, let us consider finer levels of the laser and the detection screen (illustrated in Figure 5.2). The part of the laser that emits is a collection of excited atoms, while the detection screen is a collection of atoms in their ground states. Recall that an excited atom is an atom whose electrons are in some higher-energy state than their lowest-energy (ground) state. Therefore, they could emit a photon and still have a lower stable final state to retire to. Thus, when we consider the finer details of the macroscopic emitter and absorber, we are dealing with composite quantum objects; i.e., bound states, specifically atoms. These atoms define both endpoints of the photon transaction. That is, as we noted earlier, the atoms are the microscopic emitters and absorbers. The microscopic emitters and absorbers are just quantum objects, and therefore they are not spacetime objects; they exist only in Quantumland. But, in view of the localizing effect of an actualized transaction, the microscopic emitter and absorber become localized when participating in an actualized transaction. At this point, they are still microscopic objects in the sense that they are very small, but they have become restricted to a particular region of spacetime and their overall behavior could therefore be described accurately by classical physics.[11]

One detail needs to be filled in here on a still finer level to complete our discussion of bound states. While the emitting and absorbing atoms are certainly participating in the actualized transaction, the entities actually losing and gaining the photon's energy are the electrons in the atoms. So, for example, let's consider an excited hydrogen atom (call it E) and another hydrogen atom in its ground state (call it G). Suppose an actualized transaction occurs between these two, so that a photon is given up by the electron in E and gained by the electron in G. The transaction has been actualized between these two electrons, but they are components of the bound states (that is, the atoms) E and G. So what gets localized in each case is the entire bound state. That is, when G's electron absorbs the photon and moves up to an excited state, G', we don't just have an excited electron, we have an entire excited bound state (the atom G'). Similarly, when E's electron gives up the photon, we have an entire ground bound state, E'. The basic point here is that, in general, it is a composite object (the bound state) that gets localized as an emitter or absorber in an actualized transaction, even though one can point to an individual, non-composite quantum, such as an electron, that is actually doing the emitting or absorbing.

The Mesoscopic World

We said above that TI provides a well-defined transition zone between the microscopic (quantum) and macroscopic (classical) realms. That transition zone is called the mesoscopic level. This is an area in between the

---

[11]Atoms localized by actualized transactions are verifiably small because both our theories and our experiments tell us their effective sizes. For example, the size of the hydrogen atom in its ground state is given theoretically by the average distance of the electron from the nucleus, which is roughly half an Angstrom (a very tiny unit of distance). This can be corroborated by measurements.

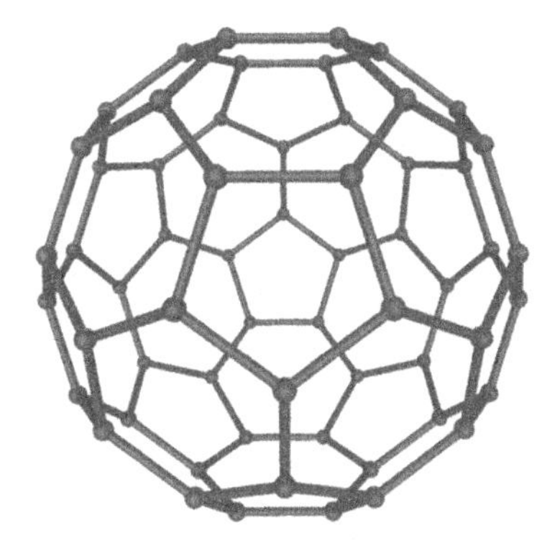

Figure 5.3. A drawing of a Buckeyball.[12]

obviously microscopic (such as a single atom) and the obviously macroscopic (such as a detector screen). A mesoscopic object is a very large bound state comprised of many atoms or molecules. An example of a mesoscopic object is a 'Buckeyball' (Figure 5.3).

A Buckeyball is a large molecule composed of 60 carbon atoms, arranged in pentagons. These objects are named after Buckminster Fuller, who invented the very stable 'geodesic' structure that supports this molecule. Mesoscopic objects like this may be composed of a rather large number of atoms, but they still need to be described by quantum mechanics rather than classical physics.

For example, one can do a two-slit experiment with Buckeyballs and get an interference pattern, which means that there is no fact of the matter about which slit it went through. This might seem surprising because of the size and complexity of the Buckeyball, but we must remember that since it does not reliably emit or absorb, it is in Quantumland and not in the spacetime realm. This means that it can propagate through the slits without necessarily triggering a 'which slit' confirmation.[13] In this case,

[12]GNU Free Documentation License. Image obtained from http://upload.wikimedia.org/wikipedia/commons/4/41/C60a.png

[13]In Chapter 3, we had 'which slit' photon confirmations from telescopes. In the case of a more complex object with mass (such as a molecule), a 'which slit' confirmation can be generated by interactions between the object and the material making up the slits. Another way to get a 'which slit' confirmation is by bouncing photons off the object. That situation is discussed in Kastner (2012, §3.3).

the Buckeyball remains in a superposition of 'which slit' states until it interacts with the final screen. At that point, one of the atoms in the Buckeyball emits a photon offer which receives a confirmation from one of the absorbing atoms in the detection screen, and it is only at that point on the final detection screen that the Buckeyball is localized as a spacetime object. Since there is no fact of the matter as to which slit it went through, an interference pattern can be seen.

Thus, objects such as Buckeyballs, despite their complexity and effective size, can have a wave nature that behaves like much smaller quanta, such as electrons and photons. The behavior of quantum possibilities is therefore apparent even into the mesoscopic realm.

Revisiting Schrödinger's Cat

The transactional picture explains why it is not only unnecessary to describe a Geiger counter or a cat by a quantum state, but it is also inaccurate.[14] Recall from Chapters 2 and 3 that a quantum state is a possibility offer. Once the offer receives a confirming response (or a set of confirming responses), the system undergoes collapse, yielding a definite result. This situation is no longer correctly described by a possibility offer, but by a classical 'brick,' as discussed in Chapter 2. The brick corresponds to a spacetime event. If what you have is a spacetime event, it is not correct to describe it by a quantum state. These are two fundamentally different kinds of entities.

Now that we've discussed the relativistic domain, with its concept of coupling, we can treat the Schrödinger's Cat riddle more accurately. Within the nonrelativistic picture, which lacks this kind of coupling, quantum theory has to model the state of the unstable atomic nucleus as a

[14]Technically, one can apply an approximate quantum state to an object of any size and complexity. But that state will not be an offer wave. For example, one can describe a large molecule by a 'center of mass' position state, but that is not an offer wave, it is an approximate description. The molecule is actually detected not through a 'molecule confirmation wave' but through transactions between its constituents and the detection screen. Also, if we are talking about a truly macroscopic object such as a baseball, the 'approximate quantum state' becomes a very poor approximation, because we have a conglomerate of actualized transactions, not a real quantum state.

superposition of decayed and undecayed quantum possibility triangles. But with the sharper tools of the relativistic theory, it turns out that the atom's nucleus is actually best understood (in the transactional picture) as a microscopic emitter candidate with an amplitude to emit an electron offer wave. Just as an electron has an amplitude to emit a photon, the neutrons in the nucleus have an amplitude to emit an electron. This is the basis of radioactivity.[15] (Remember that this amplitude squared is the probability that a quantum will be emitted and be absorbed.) However, it cannot be relied upon to emit with certainty at any given time.

Thus, the atom's nucleus can be described in the relativistic transactional picture as a candidate emitter of electron possibility offers. However, it's a special kind of candidate emitter, in that its candidacy increases in potency over time.[16] This is a characteristic of unstable radioactive atoms, and is expressed in the atom's half-life, which is the period of time at which half of a sample of that element has decayed. The Schrödinger's Cat scenario assumes that we're working with a radioactive atom whose half-life is about 1 hour.

The atom's emission candidacy is directly analogous to the coupling amplitude, discussed earlier in connection with electrons emitting photons. In the Schrödinger's Cat scenario, the atom's effective coupling amplitude, or tendency to emit, strengthens during the hour of the experiment to a point where it has a probability of 50% of emitting an electron possibility offer by the final minutes of the experiment. (Its variability comes from the other factors mentioned earlier in connection with the energy conservation hurdle concerning the availability of stable final states available to the system doing the emitting.)

Let's consider the experiment from the moment the components are put in place and the box is closed. At any particular moment, the atom's nucleus has an amplitude to emit an electron possibility offer, and that

---

[15]This happens by way of a kind of interaction called the 'weak' nuclear interaction, and involves another kind of field quantum as an intermediary. These details need not concern us here.

[16]Actually, an ordinary excited atom also has a time-dependent amplitude for emission of photon offers. The time dependence of the emission enters when the various energy conservation conditions are taken into account. These conditions result in a decay rate, and that in turn yields a time dependence.

tendency increases over time. If there are enough candidate absorbers available to it, at least one of them will respond to any such offer with a confirmation. By saying that candidate absorbers are available, we mean that there are objects that can couple to the emitted offers and provide a stable final state for themselves and the emitting object. The Geiger counter has just such entities: gas molecules that are very easily excited (by absorption) to the point where they will produce a detectable electric current.[17] An incoming electron offer wave emitted from a decaying nucleus can excite one of the electrons in the gas molecules by engaging in a photon transaction with it.

The availability of reliable absorbers (e.g., the Geiger counter) will create a fact of the matter, at any time, as to whether there is a confirmed offer or not. That is, there may be virtual electron exchanges between the unstable atom and the absorbing atoms in the Geiger counter, but these are neither emitted nor confirmed interactions, so they do not set up any incipient transactions that could be actualized, and therefore do not transfer any energy. The atom will not decay unless there is a definite emission of an offer wave and a confirmation from at least one of the absorbing atoms. These two processes set up an incipient transaction that can be actualized and thereby transfer real energy from the unstable atom to the absorber. Such an occurrence is inherently unpredictable, but it is characterized by a well-defined probability (the square of the effective coupling amplitude). That means that there is always a fact of the matter as to whether the atom has decayed or not. There is no need for a superposition of the Geiger counter, the cat, or the experimenter. There is no need for an outside observer to 'collapse' the state of the cat to alive or dead.

[17]The scattering by an incoming electron offer of an electron in a gas molecule sets off a chain reaction in which many of the gas molecules emit electrons (become ionized). This cascade effect results in a macroscopic electric current that can easily be detected. One way that the current is detected is by the fact that it generates a magnetic field. At the quantum level, such fields are propagated by transactions that involve photons (the carrier of the electromagnetic field). Thus, even when it is electrons that are emitted, the transactions that provide for collapse often involve photons. Readers interested in the quantum basis for classical electromagnetic fields may wish to look into 'coherent states' or 'Glauber states.' (See, e.g., Kastner, 2012, §6.5). For the purposes of this book, such classical fields are actually ongoing transactions involving the emission and absorption of huge numbers of individual photons.

We can now clearly see what is wrong with the usual approaches to quantum theory that do not recognize absorption. Without absorption, all we have are processes described by amplitudes. When a process is described only by an amplitude, as opposed to a probability, there is no fact of the matter as to whether it occurs or not; it is just a quantum possibility. So in the case of Schrödinger's Cat, if we neglect absorption, instead of a probability of decay we have only an amplitude. The problem of Schrödinger's Cat rears its ugly head precisely because the presence and participation of absorbers is neglected. If we neglect the absorbers, the nucleus is erroneously considered as an isolated object which only has an amplitude to emit an electron. An amplitude is not a well-defined probability, so it cannot correspond to any spacetime event.

However, with the aid of the transactional picture, we have a specific physical reason for why we are squaring the amplitude, and why the macroscopic objects are not in quantum superpositions. Because of the availability of absorbers, there is a well-defined probability at any time of whether the atom has sent out an electron offer and that offer has received a set of confirmations,[18] resulting in the decay of the atom.

Summing Up

In this chapter, we saw how the transactional picture provides for a connection between the relativistic and nonrelativistic theories. It does this by locating the birthplace of offer waves ('possibility triangles') and their responding confirmations in the basic tendency of quantum fields to couple; i.e., to emit and absorb quanta of other kinds of fields. This coupling tendency can only be described accurately by relativistic quantum theory. Coupling appears in the standard relativistic theory as the exchange of virtual quanta, as discussed in Chapter 4. However, according to the transactional picture, coupling is also the basis for the creation of offer waves under the right conditions. When that happens, the offer waves generate confirmations, and we are back to the 'bow ties' of Chapter 3. The domain

[18] I'm saying a 'set of confirmations' here because the offer wave is emitted as a spherical wave that has many components reaching different absorbers, each of which responds with a confirmation corresponding only to the component it receives. In addition, the probability referred to here is simply the decay rate in the conventional theory.

of possibility offers and incipient transactions can be well described by the nonrelativistic version of the transactional picture. At the nonrelativistic level, we can ignore, for practical purposes, the existence of virtual quanta.

What are the implications for the world of our experience? Everything we see and feel is made knowable to us in the usual empirical sense because of transactions through our senses of sight, touch, taste, sound, or smell. And the things made known are not solid little atomic particles. They are quantum possibilities that are making their presence concrete through transactions. It is the transactions that actualize certain possibilities as concrete events in our spacetime realm of sensory experience. Many others are not made manifest; there are too many possibilities for all of them to be made concrete at once. Reality is fundamentally Quantumland, and the spacetime realm that we experience with our five senses is really only the tip of the iceberg.

The Hindu term *maya*, meaning 'illusion,' could be seen as describing the spacetime realm that we experience, in that it seems so concrete and substantial, but is not the fundamental reality. This echoes Plato's view that the world of appearance is a kind of imperfect or partial version of the fundamental reality. We will examine these and related ideas in the next chapter.

Chapter 6

Reality, Seen and Unseen

'I saw... that the Light consisted of innumerable Powers and had come to be an ordered world, but a world without the bounds of material existence.'

Hermes Mercurius Trismegistus, The Divine Pymander
(translated by John Everard)

In this chapter, we're going to consider some traditional philosophical questions about the nature of reality, and see how the transactional picture can shed new light on these questions. So far, we've considered the distinction between the macroscopic world and the microscopic world. We've seen that classical physics describes the macroscopic world quite well, but that it fails to adequately describe the microscopic world of atoms and other fundamental components of matter. We've also noted that the world of appearance corresponds to the macroscopic realm describable by classical physics, while the underlying, hidden reality can be seen as corresponding to the quantum level. It turns out that this distinction between the macroscopic and microscopic realms, and the need for different physical theories to describe them, is directly relevant to some longstanding philosophical puzzles.

The Tree in the Forest: Realism vs. Antirealism

Here's a famous question that illustrates the kind of puzzle we're going to consider: If a tree falls in the forest, and there is nobody there to hear it, does it make a sound? If you think that it does make a sound, and — more generally — that such an event does not require any conscious, perceiving observer in order to be considered real, then your view is described in philosophical terms as realism. Realists hold to the seemingly common-sense idea that objects and events exist independently of subjective perception. So, for example, a realist would say that the teacup

I'm holding in my hand does not depend on my perceiving it in order to exist. If it seems strange to you to consider otherwise, then you are in good company: most of us start out life as realists. It seems like a very natural view. Indeed, it seems distinctly odd to suppose that a teacup does not exist without someone perceiving it, or that when a tree falls it does not make any sound unless someone hears it.

Nevertheless, the opposite conclusion about reality has been arrived at by many philosophers who have carefully considered questions like this. The opposing view, that objects and events don't exist unless someone perceives them, is called *antirealism*. The antirealist believes that an object is constituted by no more than an observer's subjective impressions. In other words, the antirealist asserts that there is nothing more to the object than that set of perceptions. How might one arrive at this view? One way is by questioning whether we can ever get outside our own private mental world to confirm that there really is anything more to an object than what we can perceive.

A graphic example of this challenge is found in the movie *The Matrix*, where the characters are living in an illusory world created by the electrical stimulation of their brains by hostile intelligent machines. We first encounter the main character, Neo (played by Keanu Reeves), in what appears to be an ordinary world. He has a job as a computer programmer, and goes to restaurants and nightclubs. He enjoys the noodles at his favorite restaurant and spends his free time hacking into various computer systems. But as the movie unfolds, we find that none of those perceptions have corresponded to independently-existing objects and events. In fact, Neo has been plugged into a computer program, the Matrix, which has artificially created all of his experiences by stimulating the appropriate perceptual areas of his brain.

The antirealist view is based on taking this sort of scenario seriously: all we can really know about for sure is what we perceive. We like to think that what we perceive is being caused by something that exists objectively, outside our subjective perceptions, and that does seem like a reasonable assumption. For example, we normally assume that when we're holding a teacup, there is some objective physical object out there that looks just like a teacup and that is causing our perception of the teacup. But we cannot verify, through sensory observation, that there really is such an object outside the perceptions in our minds, simply because we cannot get outside our own minds. Therefore, the antirealist rejects the notion of the independently-existing, objective teacup as an unsupportable assertion. The antirealist concludes that the only sense in which any object or event exists is in a subjective impression in one's mind. That is, the antirealist says that any object or event is constituted by the subjective impressions of someone who perceives it, and nothing else. A famous antirealist, the Irish philosopher George Berkeley, put it this way: 'To be is to be perceived.'

At this point, however, it should be noted that while one cannot prove that there *are* objectively-existing entities outside our minds, one certainly cannot prove that such entities do *not* exist. This book presents a case that there are very good reasons to believe that there are independently-existing entities — quantum systems — and that assuming that these are physically real provides a natural way to understand why quantum theory is so empirically successful and yet so resistant to interpretation in the usual way.

We can express the opposing views of realism vs. antirealism in terms of the *subject/object distinction*. The object is what we wish to know about, and the subject is us; or, more precisely, the aspect of our minds that acquires knowledge. We generally take it for granted that we, as subjects, perceive objects as they exist independently of our observations. And indeed, science has always presupposed that there is a clear distinction between an object of study and the person (subject) studying it. In this traditional, commonly-held view, the object determines what our subjective impressions will be, and that's how our knowledge of the object is acquired. Another way to put this is that the object itself drives our knowledge of it.

The antirealist, however, denies that there are independently-existing objects out there to be passively known. Instead, he or she believes that all knowledge is subject-driven. If you think again about *The Matrix*, you can see an example of this: Neo's 'knowledge' of objects and events in the Matrix were entirely subject-driven; that is, all those phenomena were constructed in his mind. There were no independently-existing office buildings, restaurants, or night clubs.[1]

---

[1]Readers familiar with Descartes' *Meditations on First Philosophy* (1641) will recognize this is the 'evil demon' scenario, in which an evil demon is deceiving us by having us perceive things that are not there at all. The updated version of the evil demon scenario is the 'brain in a vat' scenario, in which a disembodied brain is artificially stimulated to perceive that he/she is a person experiencing an ordinary life. The people plugged into the Matrix exemplify this updated version.

Rationality and Knowledge

I said above that 'all we can really know about for sure is what we perceive,' where perception means sensory perception. But is this strictly true? Do you think you could assert that 2 + 2 = 4 without necessarily having any sense perception? If so, then you also believe in the power of rational thought to yield knowledge. The view that people can gain knowledge by using rational thought alone is called *rationalism*.[2] On the other hand, some researchers disagree. They argue that the only way one can obtain knowledge is through observation, using our five senses. This view is called *empiricism*. An empiricist holds to the dictum: 'I'll believe it when I see it, and not before.' While an empiricist can accept that 2 + 2 = 4, he would argue that you arrived at that idea by observation, not by rational thought alone.

Physical science offers an interesting amalgam of these two approaches to knowledge, since it certainly requires empirical observation (i.e., experiment) but also makes considerable use of rational thought in order to develop highly mathematical and logically rigorous theories about the world of experience. (To clarify a matter of terminology: 'empirical' just means information that is obtained from observation. 'Empiricism' is the doctrine, discussed above, asserting that only observation can provide knowledge.) The brilliant physicist Eugene Wigner (1960) wrote a famous article remarking how astonishing it was that nature was apparently subject to rational analysis, and concluded with this comment: 'The miracle of the appropriateness of the language of mathematics for the formulation of the laws of physics is a wonderful gift which we neither understand nor deserve.'[3] Physicist Freeman Dyson made a similar comment, quoted in Chapter 1, concerning the seemingly serendipitous applicability of complex numbers to the real world.

The basic point is this: when we combine careful observation and imaginative rational theorizing, we can often arrive at amazingly accurate

---

[2]A strict empiricist would not deny that we can find logical truths; he would, however, assert that we got them through sense perception, not by rational thought alone.

[3]'The Unreasonable Effectiveness of Mathematics in the Natural Sciences,' in Communications in Pure and Applied Mathematics, vol. 13, No. I (February 1960). New York: John Wiley & Sons, Inc. Copyright © 1960 by John Wiley & Sons, Inc.

theories about the phenomena we observe. By 'accurate theories,' I mean that we can use these theories to successfully predict what will happen in certain circumstances. Unlike the empiricist or the antirealist, the realist can readily explain the predictive success of such theories by supposing that the theories describe real entities and processes, even if we can't necessarily perceive those entities and processes. Early atomic theory is a case in point, so let's take a moment to consider a relevant bit of history.

It may surprise the reader to learn that less than 200 years ago, most physicists took a dim view of the idea that there were tiny, unseen things called 'atoms' that served as the fundamental constituents of material objects. That was thought to be a quaint idea from the ancient Greeks, and nothing more. But the person who took the idea of atoms seriously, Ludwig Boltzmann, revolutionized physics by showing that if he assumed that atoms really existed and mathematically analyzed their behavior, he was able to accurately predict the observable behavior of macroscopic samples of matter.

However, Boltzmann's work was not well received in his day. His staunchest critic was the empiricist Ernst Mach, who viewed atoms as a scientifically-unsound fiction. According to Mach, physics should stick to formulating laws based only on observations. Examples of such laws are the various laws of thermodynamics, which describe relationships between observable properties of matter such as temperature, pressure, and volume. Recalling our definition of the empiricist, Mach would only 'believe it if he saw it,' and atoms were theoretically-proposed objects that were not directly observable. So Mach did not believe that they existed, and neither did most other physicists of the period. This general lack of acceptance of Boltzmann's approach eventually led to his falling into a deep depression, and in 1906 he took his own life. Ironically, this was six years after Planck used Boltzmann's ideas as a crucial ingredient for successfully deriving the correct black body radiation law,[4] and one year after Einstein, in turn, used Planck's work to successfully account for the

[4]Boltzmann's statistical analysis of atomic behavior helped to explain why it is harder to excite standing waves in a black body (modeled as a black box) with larger-sized energy packets. The details are beyond the scope of this book, but the reader is invited to consult Eisberg and Resnick (1974, §1–4) for a clear account.

photoelectric effect. Moreover, Einstein would go on to successfully apply Boltzmann's ideas to other challenges in atomic physics. It's probably not an overstatement to say that Boltzmann's imaginative analysis of his unobservable atoms paved the way for the crucial breakthroughs of quantum theory.

The vindication of Boltzmann's atomic approach lends support to the idea that observable phenomena can indeed be explained by the existence and behavior of unobservable, or at least thus-far-unobserved, entities. However, even though some of Boltzmann's mathematical approaches were used to formulate quantum theory, it has been far from clear what quantum theory is 'about.' That is, what exactly is the theory describing?[5] One difficulty in coming up with a physical subject matter for quantum theory — that is, a description of what it is that quantum theory is 'about' — is that the very nature of what exists seems to depend on the act of observation; that is, on measurement. Recall that the antirealist holds that what exists is constituted by subjective perceptions, not by any unobserved entity or substance 'underneath' those perceptions. Thus, many researchers have adopted antirealist views about reality based on quantum theory, meaning that they deny that the theory describes anything real. They say instead that it describes only our knowledge.

I believe that it is a mistake to conclude that quantum mechanics implies that we should be antirealists. The mistaken conclusion arises from two interpretive failures concerning quantum theory: (1) under traditional approaches to the theory, the concept of measurement has not been adequately defined in physical terms, and therefore has been viewed as observer-dependent and subjective when it is not; and (2) it has been unnecessarily assumed that 'to exist' means 'to exist within spacetime.' Regarding factor (1), this is just the measurement problem. We have already seen in Chapters 3 and 4 that the transactional picture can give an unambiguous, physical account of how 'measurement' occurs, and the process is not dependent on an observing consciousness. It simply

[5]Of course, quantum theory is about atoms too; but Boltzmann's classical model treated them as tiny billiard balls (which worked for his purposes of deriving the macroscopic laws of thermodynamics). In contrast, quantum entities cannot be treated as tiny billiard balls. Hence there is a special mystery facing quantum theory.

requires emitters and absorbers, which are ubiquitous. These processes are physical in nature, and not based on subjective perceptions, so the transactional account of measurement avoids the slide into antirealism.

Factor (2) is the philosophical mistake of assuming that the only things that can exist are those that are within one's world of appearance. This is the same error made by our Square in Flatland when he assumed that the Sphere had to be lying about his own three-dimensional world. That is, because the Square's world of appearance contained only two dimensions, he ruled out the existence of any higher-dimensional entity. It is also the same error made by the prisoners in Plato's Cave when they assume that all that exists is the shadow play on the wall. And it's the same error made by Ernst Mach in rejecting the existence of atoms on the basis that they were unobservable. The moral of the story is that even very clever people, like Mach, can make interpretational mistakes.

Is there Common Ground for the Realist and the Antirealist?

The transactional interpretation of quantum mechanics can provide an interesting and significant area of common ground upon which the realist and antirealist can meet. It's in the form of a compromise: there is something that each party will like, and something that each will find less desirable. The compromise offered by this proposed interpretation of quantum theory is as follows: the world is more observer-dependent than the die-hard realist might think, but less observer-dependent than assumed by the antirealist. We'll examine this common ground first in terms of the work of the famous 18th-century German philosopher Immanuel Kant.[6]

Immanuel Kant's work can be seen as an interesting elaboration of Plato's basic idea that there are two aspects of reality: (1) the world of appearance and (2) the underlying reality. Kant provided his own terms for these two aspects of reality. He called them (1) the phenomenal realm and (2) the noumenal realm. The word 'phenomenon' has etymological roots in both the Latin and Greek words meaning 'appearance.' So the term 'phenomenal realm' has the same meaning as 'the world of appearance.' It specifically refers to sense perception; that is, perceptions arising

[6]Relevant work is contained in Kant (1966).

from one or more of the five senses. Earlier in this book, we noted that the phenomenal realm is basically the spacetime realm.

In contrast, the term 'noumenal realm' refers to that aspect of reality which is not knowable by way of the five senses, but which may be known about through the intellect or other non-sensory means. Kant used the term 'noumenon' to mean that hidden, unseen aspect of an object that is not perceivable through the five senses. (The plural form of this term is 'noumena.') So the phenomenal realm is the realm of phenomena — appearances — and the noumenal realm is the world of noumena, things as they are underneath their appearances. Kant also referred to noumena as 'things-in-themselves.' By this terminology, he was trying to indicate that the noumenon is that objectively-real aspect of an object that doesn't need to be perceived in order to exist.

It's important to note that an antirealist would deny that there is any such thing as a 'noumenal realm,' since antirealism holds that knowable objects are constituted by our sensory perceptions and nothing else. So an antirealist denies that there are 'things-in-themselves'; that is, he or she denies that there is any aspect of an object that doesn't need to be perceived in order to exist. The antirealist thinks that the only way in which any object exists is through its being perceived.

Above, we considered Kant's ideas in the context of Plato's formulation. This was a realist formulation: the realist acknowledges that objects present appearances, but he or she also thinks that the objects that we perceive exist independently of our sense perceptions. That independent existence of the object is the unseen reality of the object, the noumenon. In contrast, the antirealist thinks that objects only exist insofar as we perceive them; according to the antirealist, there is no 'unseen reality.' In terms of the subject/object distinction discussed above, the realist thinks that knowledge is about an object, while the antirealist thinks that knowledge is constructed by a perceiving subject.

To gain further insight into this disagreement, consider a famous discussion by Bertrand Russell in his classic book, *The Problems of Philosophy* (1959). Russell starts by considering an ordinary table. He notes that we like to think of the objectively 'real' table as existing independently of our perceptions of it, but then argues that it's very hard to specify what that 'real' table is. This is because there are many different points of view, scales of magnification, degrees and types of light, etc., under which one could view the table, and these views will not necessarily agree with one another. For example, the table may appear smooth and shiny to the eye, but rough and textured under a microscope. The table's appearance always depends on the conditions under which it is perceived. Russell famously concludes that the only knowledge we can have of the table is of various aspects of its appearance; the 'real' table underneath the appearances — whatever that might be — is a deeply-mysterious object.

According to the realist, all the different views are subjective, observer-dependent impressions of the real table, but not the real, objective 'table-in-itself.' The realist, nevertheless, maintains that there is a 'real table' — that is, a noumenal table, in Kant's terms — somewhere out there, independent of our perceptions of it. But the realist encounters great difficulty in specifying what it is. Meanwhile, the antirealist denies that there is a 'table-in-itself' at all. According to the antirealist, each of us perceives a different table, and that's all that exists; for antirealist, there is simply no 'noumenal' realm.

However, we can indeed find a common ground between these two views, courtesy of the transactional picture of quantum theory. It allows the realist to have the 'things-in-themselves,' or noumena, but the price paid by the realist is that these noumena are very different sorts of entities from their appearances. That is, the table-in-itself does not look like a table; it is much less concrete. Meanwhile, the antirealist is granted that the phenomenal objects existing in the spacetime realm are indeed dependent on observation, in a way: they are dependent on measurement. However, the price paid by the antirealist is that the phenomenal objects are not wholly observer-dependent and subjective, as he had assumed. This is because in the transactional picture, 'measurement' is a physical process, namely absorption. The confirmation of an offer wave and ensuing transaction really happens, whether or not the result happens to be observed by a conscious entity. So that which exists is perhaps less objective than the realist believes, but more objective than the antirealist believes.

Here's how the transactional picture of the quantum realm makes the compromise work. The unseen noumenal realm is composed of quantum systems, which are intrinsically unobservable. However, despite being unobservable through the senses, they can certainly give rise to sensory perceptions in the phenomenal realm of spacetime under suitable conditions, such as 'measurement.' In the transactional picture, 'measurement' just means that absorbers respond with confirmations. The confirming response of the absorbers occurs whether or not a subjective consciousness is involved; that response is an objective, physical event, as is any actualized transaction arising from the confirmation. So the absorption takes the place of the requirement that in order for something to exist it must be 'perceived' in the subjective consciousness of an observer. That is the good news for the realist. The 'bad' news for the traditional realist is that the offer wave alone does not correspond to any object or event in spacetime; that is why it is a noumenal entity.[7] In order for an object or event to become manifest in spacetime, i.e., to become a phenomenon, an absorber response is required. So any object or event that can be known in spacetime is dependent on the manner in which its underlying offer wave(s) interact with absorbing systems, among which are our sense organs. Figure 6.1 depicts this process.

These sensory experiences — the set of phenomena comprising the phenomenal table accessible to the man in Figure 6.1 — correspond to specific actualized transactions. But the table-in-itself is just a collection of interacting quantum systems. These systems are capable of scattering

Object ⇔ Offer wave

Subject ⇔ Confirmation wave

Phenomenon ⇔ Transaction

Figure 6.1. Transactions create the objects-of-appearance. The man is observing a table using sight and touch.

[7] And this is why 'local hidden variables' theories do not work. Such theories are attempts to describe quantum objects in spacetime terms — but spacetime is the phenomenal realm, and quantum objects live in the noumenal realm.

light (photons), and in that sense serve as emitters of offer waves.[8] If we are observing the table, our eyes constitute the relevant absorber systems that respond to offer wave components with confirmations. On a macroscopic scale, there are enormous numbers of such offers and confirmations. These interactions set up enormous numbers of incipient transactions, many of which will be actualized. Those, in turn, deliver real energy from the quanta comprising the table-in-itself to our eyes, and each such energy transfer creates an aspect of the phenomenal table that we will perceive.[9]

In terms of the subject/object distinction, the independently-existing object is the set of quantum systems comprising it, which can emit offer waves under suitable conditions. The subject is the absorber configuration, which determines the confirming responses to those offer waves. The spacetime phenomenon is the actualized transaction between any particular emitter and absorber. Thus, each observer will experience a different phenomenal table because each will experience a different set of actualized transactions. But no conscious observer is required in order for there to be objectively-existing transactions that deliver real energy from one entity to another. So the term 'subject' in this compromise does not necessarily signify someone's mind, as the antirealist would prefer. As noted above, it simply refers to the forces and absorber configurations acting on one or more offer waves.[10]

In the next section, we'll consider an example that illustrates how noumena can be very different from phenomena, despite the fact that they are indeed the 'things in themselves' that give rise to those phenomena.

---

[8]Technically, an object such as a table scatters offer waves from another emitter, such as a light bulb. But the scattering will yield information about the table, so for present purposes we can think of it as emitted by the table.

[9]Of course, our eyes then transmit signals to our brain. This is just another transaction. The question of how this physical process can lead to subjective experience is a fascinating one and a fertile ground for further exploration. Is the mind a quantum entity? The present interpretation is open to a possible place for the mental, subjective realm in Quantumland, even though it does not *require* subjectivity in order to solve the measurement problem.

[10]One might wonder whether quantum entities are endowed with some rudimentary form of consciousness. Such considerations are interesting and important, but beyond the scope of this book.

Games and Reality[11]

We can gain some insight into the idea of 'things in themselves' by considering an example from popular culture: online role-playing games. Many readers will be familiar with types of online games called *massively multiplayer online role-playing games* (or MMORPGs for short). Examples of these are *Second Life, World of Warcraft*, and *Eve Online*. In these games, a player, or 'user,' accesses the online game environment by loading a software package onto his or her computer. The software enables the player to create a character, or 'avatar,' which represents him or her in the online game environment. Let's consider two users, Jonathan and Maria. Jonathan's game avatar is called 'Jon,' and Maria's is 'Mia.' Once the avatars are created in the game environment, they carry with them an individual point of view (POV). Each user can monitor what his or her avatar perceives through this POV as the avatar pursues its in-game career.

One of the many tasks that the user Jonathan can have his avatar Jon engage in is to create objects in the game environment. For example, Jonathan may have Jon create a table. To do this, Jonathan inputs the required commands through his avatar Jon into the game environment, and a 'table' appears at the desired 'location' in Jon's vicinity. Note that I've put quotes around 'table' and 'location' because these are not a real table or a real location in an objective sense. They are part of a phenomenal environment; that is, avatar Jon's world of appearance.

Now suppose Maria is playing the same game; recall that her avatar is called 'Mia.' Maria might be sitting at her computer in Madrid, while Jonathan is in Stockholm. Nevertheless, their avatars may be in the same game environment 'room'; let's say the 'Philosophy Library.' Now (just for the sake of our analogy) let's pretend that the avatars, Jon and Mia, have some self-awareness. But suppose they don't know that they are only avatars, and instead think of themselves as autonomous beings (very much like Neo's 'Mr. Anderson' and the other plugged-in human subjects in *The Matrix*.) Let's imagine Jon and Mia discussing the table in front of them, along the same lines as the discussion in Bertrand Russell's *The Problems of Philosophy*. Recall that Russell argues that the appearance of the table

[11]This section is based loosely on Kastner (2012), §7.5.

depends, to a great extent, on the different conditions under which it is perceived, and that we have no direct knowledge of the real table:

> Thus it becomes evident that the real table, if there is one, is not the same as what we immediately experience by sight or touch or hearing. The real table, if there is one, is not immediately known to us at all, but must be an inference from what is immediately known. Hence, two very difficult questions at once arise; namely, (1) Is there a real table at all? (2) If so, what sort of object can it be? (Russell, 1959, p. 11)

Russell's presentation is an account of the deep divide between, in Kant's terms, the world of appearance (phenomenon) and the thing-in-itself (noumenon). (Notice how he repeats the phrase 'if there is one,' to emphasize how little we really know about it.)

If avatars Jon and Mia were to pursue this analysis, they, too, would find that the only knowledge they have of the table is based on its appearance (which their human users can monitor on their computer screens showing their avatars' POVs). Suppose the table presents two apparently conflicting properties: the side of the table first facing avatar Jon is two meters long, but other side, facing avatar Mia, is only $1\frac{1}{2}$ meters long. They might at first disagree on the size of the table, but they could discuss what they see, and could agree to compare their perceptions by, say, changing places. Mia could then confirm that Jon's side of the table is longer, and vice versa. They would realize that each of them is seeing the table from a different vantage point, and it is really a trapezoid. By performing these sorts of comparative observations, Mia and Jon could convince themselves that there really is a table there because they would be able to corroborate their different perceptions in a consistent way: their public observations would form a consistent set. This in turn would suggest to them that there is something out there that is the direct cause of their perceptions. In traditional realist fashion, they might conclude that there is a real, but unseen, table behind or underneath the appearances — a 'table-in-itself' — that causes and resembles their perceptions of it.[12]

[12]The naïve realist notion that independently existing objects outside the mind are the causes of ideas (perceptions) that resemble them is examined in Descartes' *Meditations* (1641). Descartes ends up deciding that this is more or less right, but his solution remains subject to debate, as does naïve realism.

But what about the human users, Jonathan and Maria? They both know that, in some sense, there is a 'table-in-itself' that could be said to be the cause of Jon and Mia's perceptions of the game table. But the 'table-in-itself' does not resemble the game table at all. What is the 'table-in-itself'? It is nothing more than binary data (on a server somewhere at some other location), manipulated by the people who created the game and by the human users (Jonathan and Maria). Compared to the game table perceived by the avatars Jon and Mia, it is insubstantial, abstract. And yet, clearly, it is the direct cause of the avatars' perceptions of an ordinary table; the 'table-of-appearance,' or phenomenal table. To them, this phenomenal table is certainly not just an 'illusion': the avatars cannot ignore it. If, for example, they were to try to run through it as if it weren't really there, they would bump into it and may even incur physical damage (as measured by their 'health' levels in the game environment). So their interactions with the phenomenal table have real, physical consequences at their level of experience.

If a human user were to somehow speak to an avatar like Mia and tell her that the objects in her world are nothing but information, she would scoff at the suggestion, and she might ask why she suffers damage if she falls off a cliff in her 'only information' world. To the avatars, their phenomenal world — the game environment — is concrete and consequential. Meanwhile, in this analogy, the users Jonathan and Maria and their data-based manipulations of the game are taking place in the noumenal, unseen realm.

What does this little parable tell us about our world of ordinary objects-of-appearance; that is, our phenomenal, empirical world? It tells us that it is conceivable, and even quite possible, that the 'table-in-itself' of our world is a very different entity from what the table-of-appearance might suggest.[13] We, and the objects around us, are governed by the laws of physics (the 'rules' of our game). So we interact with these objects and are affected by them, and in that sense they are certainly real and consequential, just as the game-environment objects are real and consequential for

[13]This case is different from the 'evil demon' or 'brain in a vat' scenario. There is no deception involved here; the things-in-themselves are the natural causes of the observed phenomena. They are just different kinds of objects than we thought.

avatars Jon and Mia. But the 'object-in-itself,' or noumenon, is precisely *that aspect of the real object which is not perceived.* If an unperceived aspect of an object exists, we can reasonably expect it to be an entirely different entity from those in our perceived world of experience. As noted above, in the transactional picture, the 'object-in-itself' can be considered to be the offer wave(s) capable of giving rise to the transactions establishing the appearances of the object. Just as the 'table-in-itself' underlying the avatars' table does not really exist in their game world and is a kind of abstract information, so the quantum offer waves giving rise to our real empirical objects do not exist in spacetime, but nevertheless have an abstract (but physically-potent) reality in Quantumland. They are noumena.

Schrödinger's Kittens: Nonlocality Explained

We first met the Schrödinger's Kittens riddle in Chapter 2, but let's review it briefly here. This experiment, given its colorful name by author John Gribbin (1995), involves two correlated electrons that have a nonlocal influence connecting them. If these electrons fly apart to widely-separated measuring devices, those measurements will always show correlated outcomes in a way that cannot be explained by the propagation of a signal between them that has a speed equal to or slower than that of light. This experiment implies that faster-than-light influences are somehow in play at the quantum level, even though Einstein's theory of relativity tells us that no signal can propagate at speeds greater than that of light. This chapter's account of quantum objects as part of the noumenal realm, a realm beyond spacetime, provides us with a natural resolution to the apparent paradox presented by the Schrödinger's Kittens experiment. In a nutshell, the solution is that the 'cosmic speed limit' of relativity applies to the spacetime realm — the phenomenal realm — but not to the noumenal realm of quantum objects.

The Schrödinger's Kittens paradox can be visualized in terms of our game parable above. First, recall that the online game environment as observed by the avatars represents the phenomenal, spacetime realm. On the other hand, the information manipulated by the game designers and the players, Jonathan and Maria, represent the noumenal, quantum realm. Clearly, there are laws and limitations applying to the movements of the avatars and objects in the game environment that do not apply to the information and actions of human users Jonathan and Maria. For example, avatar Jon can walk around a table in his environment; indeed, if it's in his way, he must walk around it to avoid bumping into it. Also, it takes Jon a certain amount of time to walk around the table so as to avoid colliding with it. On the other hand, for human user Jonathan, the table does not function as an obstruction that he would physically need to avoid. That is, Jonathan, being outside the game environment, is not constrained by the laws of that environment.

As a preliminary, first consider the situation involving two correlated objects moving apart as they would in classical physics. Let us represent these two classical objects by the avatars Jon and Mia (see Figure 6.2). A correlation can be set up as follows: the human players Jonathan and Maria have their avatars meet in some location in the game environment. Jonathan and Maria then decide on what equipment each of the avatars will bring for a given task, thus establishing a shared purpose between those choices of equipment. For example, Jon might bring a bow and Mia might bring some arrows. Then the two avatars run away from each other in opposite directions to meet two spatially-separated equipment detectors. (These are hypothetical measuring devices analogous to a Stern–Gerlach device for measuring spin.) This correlation is a local one because it is simply propagating along with the avatars as they run from their central location; they carry the correlation along with them in a way that obeys the in-game speed limit.

Suppose that Jon will run toward a detector labeled J, and Mia will run toward a detector labeled M. Jon and Mia's speeds will be restricted to

Figure 6.2. A classical, local correlation between the avatars.

whatever is permitted by the game software, and an in-game observer (another avatar) could infer that the correlations observed in Jon and Mia's equipment were brought about in a common-sense, local manner by their initial consultation and their in-game running speeds. In particular, each avatar simply has whatever equipment they started out with. The detectors just tell us what it already was, and there is no need for any 'spooky action at a distance' as described in Chapter 2. If *J* reveals that Jon is carrying a bow, then the in-game observer can infer that *M* will detect that Mia is carrying the arrows, because that's how they started out. In our analogy, this is the 'local spacetime' account of correlations that we are accustomed to in classical physics, and detectors *J* and *M* represent classical versions of the spatially-separated measuring devices used in the Schrödinger's Kittens scenario.

Consider now the quantum situation, in which Jon and Mia represent the correlated electrons. In this case, neither Jon nor Mia has any particular kind of equipment when each departs from the central meeting place (see Figure 6.3). Instead, all they have is a sort of promissory note, if you will; an agreement that their equipment will always be properly coordinated when measured by detectors *J* and *M*. This promissory note is their quantum correlation, or 'entanglement,' and it's one that only gets fulfilled upon measurement. But there is a subtlety about those measurements. Recall that in the electron case, the observers manning each of the detectors have to choose which spin direction to measure; i.e., whether to measure 'spin in the vertical direction,' in which case the outcome would be either up or down, or 'spin in the horizontal direction,' in which case the

Figure 6.3. A quantum, nonlocal correlation between the avatars.

outcome would be either right or left. In the game analogy, the choice of type of equipment to be measured represents those spin direction choices. Suppose the in-game detectors can be set to measure either 'fire-starting equipment' (matches or fuel) or 'combat equipment' (bow or arrows). For each choice, the detector would tell you which component the avatar brought. So, for example, if the detector were set to measure combat equipment, it could tell you whether the avatar had a bow or arrows.

While in the classical case the two avatars met in a central location and their human users decided on their equipment before the avatars separated, in the quantum case the type of equipment must be chosen 'on the fly.' The avatars start at a common point together, with only their agreement that if their detectors are set for the same type of equipment, their equipment will be found to be in perfect correspondence. Then they start running apart. If both *J* and *M* are set to measure combat equipment, then seemingly miraculously, Jon and Mia will always be found to be equipped with perfectly-coordinated weaponry; one will always have a bow and the other will always have the arrows. On the other hand, if *J* and *M* are set to measure fire-starting equipment, then one of them will always have fuel and the other will always have matches. Yet they did not start with any equipment in particular, so this perfect coordination cannot have been established by their carrying the matching equipment at normal game speed through the game environment. It is as if one can send an instantaneous message to the other as soon as the first meets their detector. But such an instantaneous message would violate the in-game speed limit (which corresponds to the speed of light). In the case of a real Schrödinger's Kittens experiment, the results would be instantaneously correlated, even if the entangled electrons and their detectors were halfway across the galaxy.

So how can this be explained in the game analogy? When the avatars are near their respective detectors, their human users, Jonathan and Maria, can simply confer from their respective terminals in Madrid and Stockholm, and decide what equipment each of them is going to supply to their avatars. This outside-the-game conference by Jonathan and Maria is analogous to the nonlocal connection between the electrons in Schrödinger's Kittens that enforce the correlated measurement results. These correlations are imposed through the human users' communication channel outside the game environment (meaning outside spacetime). The correlated results from the electrons (being correlated as 'up or 'down') are likewise enforced by an influence outside spacetime.

The interesting difference between the quantum and classical case is that in the quantum case, the detectors can be set to measure two different kinds of equipment, but they will always still give a relevant outcome. That is, *J* could be set to measure fire-starting equipment and *M* could be set to measure combat equipment, and each would still give a meaningful result. For example, *J* could find that Jon was carrying fuel and *M* could find that Mia was carrying arrows. Now, of course, there is no nice match or correspondence between fuel and arrows; they do not function together as a useful set of equipment. But for measurement settings like this that do not correspond, such results occur; and even in such cases, quantum theory successfully predicts how often they will occur. In fact it successfully predicts all other possible sets of outcomes for all the possible measurement settings.

In contrast, in the classical case, it would make no sense to set *J* to measure fire-starting equipment if Jon already had some kind of combat equipment. So the basic point is that quantum systems have more flexibility as to what their properties are; they are not 'set in stone' as in the classical case, and their measurement context plays a key role in bringing about those properties. Of course, in the transactional interpretation, 'measurement context' simply means what kinds of confirmations are generated.

In our analogy, the correlation between Jon and Mia is established outside the game environment, with those binary data that seem 'abstract' and insubstantial from an in-game avatar's point of view. In the case of Schrödinger's Kittens, the correlation occurs outside our spacetime, phenomenal environment, in the noumenal, quantum realm. Thus, nonlocal influences can be naturally understood in this picture as influences communicated by noumenal entities. While these entities certainly are subject to physical laws (the laws of quantum mechanics, and conservation laws), they are not subject to limitations that apply only to spacetime phenomena, such as the restriction to speeds no greater than the speed of light.

However, it's important to note that the influences that correlate quantum entities cannot be used to send a controllable signal. Such a controllable signal would inevitably be an actualized, spacetime phenomenon (that's what it means for it to be controllable), and therefore would be subject to the cosmic speed limit. The quantum correlations are able to escape this limitation only because they operate at the noumenal level, which is the level of unactualized possibility. In the game analogy, an in-game observer (i.e., another avatar) cannot control the correlations between Jon and Mia to send 'super-game-speed signals': the correlations are controlled by the human users outside the game environment and are not accessible to any avatar in the game.

There is an obvious irony here: quantum possibilities are vastly more numerous and varied, and infinitely faster, than spacetime actualities. We can interact with those possibilities, but we cannot harness all their superior features for spacetime activities, such as communicating with observable signals, because they do not exist in spacetime. They operate strictly behind-the-scenes.

The Cheshire Cat and the World of Appearance

In Lewis Carroll's *Alice's Adventures in Wonderland* (1866), Alice encounters the famous Cheshire Cat. This mystical creature can appear as an ordinary cat, or he can choose to vanish and reveal nothing except his smile (Figure 6.4). When only his smile appears, we can think of that

Figure 6.4. The Cheshire Cat with only its grin visible.

smile as our world of appearance. Meanwhile, the rest of him that is hidden from view can be compared to the hidden reality, or the noumenal realm of Kant. In the interpretation offered in this book, this veiled part of the Cat corresponds to our hidden quantum reality. In contrast, the visible smile corresponds to the observable, spacetime phenomena we experience with our five senses. These phenomena are created and sustained through actualized transactions.

Let's consider how this phenomenal 'Cheshire Cat smile' world arises from the hidden, quantum level. Recall from the previous chapter the special status of atoms and molecules as bound states, i.e., composite quantum objects that are not simple offer waves. Much of the unseen world is comprised of these sorts of composite quantum objects. As discussed at the end of Chapter 5, these objects are the basic microscopic emitters or absorbers that can become localized in spacetime through actualized transactions. Microscopic emitters are excited atoms or molecules, which emit offer waves from their components, while microscopic absorbers are those atoms or molecules that have an available higher energy state that they could occupy if they were to receive energy. Such an absorber may respond to an offer wave of the correct energy with a confirmation, thereby setting up an incipient transaction that could be actualized and which could thereby transfer that energy. Once that happens, the emitting atom or molecule, the absorbing atom or molecule, and the transferred quantum are all localized in spacetime.[14] We can think of each of these localizing processes as painting a small part of the smile on the Cheshire Cat.

This is how atoms and molecules serve as bridges between the realm of pure offer waves (noncomposite quanta) and ordinary macroscopic phenomenal objects (such as a table). The paradigmatic pure offer wave is the photon offer wave, which serves as the intermediary between emitting and absorbing atoms. Thus (metaphorically speaking) we see the smile on the Cheshire Cat because only the atoms in its face are emitting photon offer waves (or at least reflecting them from another light source), while the

---

[14]Recall from Chapter 5 that the transferred quantum is localized only to the spacetime interval connecting the emitter and the absorber, not to a specific spacetime trajectory. The concept of a spacetime interval will be explored further in Chapter 7.

atoms in our eyes are confirming those offer waves and receiving the actualized photons (resulting from actualized transactions). The rest of the cat's body is also made up of atoms, but we can't perceive it if its constituent atoms are not participating in actualized transactions, because in that case they are not localized in spacetime. The Cat may be a highly complex object, but it cannot be perceived with the five senses unless it participates in actualized transactions with our sense organs. Similarly, much of our world is composed of complex bound quantum systems, but they are not perceivable unless they participate in actualized transactions. Our phenomenal world is only the smile on the Cheshire Cat. It is rooted in a vast, unseen, quantum world.

Chapter 7

Spacetime and Beyond[1]

'The essence of reality is being born right now. It has never existed before. Reality is constant creation and destruction, and in this constant change is something unborn and undying [...] in this space, the undiscovered and ever-changing moment exists — a moment containing all possibilities [...].'

H. E. Davey

We saw in the previous chapter how actualized transactions give rise to specific spacetime events: the emission is one event, and the absorption is another. The two are linked by the actualized quantum that is delivered from the emitter to the receiving absorber, and that link is a spacetime interval. In this chapter, we'll consider in more detail that new account of space and time. In this picture, neither space nor time exists as an independent substance or container ready to be filled with events. Instead, the transacted events themselves, together with their relationships, collectively form the structure that we call spacetime. This spacetime structure emerges, through the transactional process, from an underlying reality of possibilities: 'Quantumland.'

Appearance vs. Reality

First, let us recall Plato's two levels of reality: (1) appearance; and (2) the underlying, hidden reality. In Plato's philosophy, (1) means the world directly perceived by the five senses, and (2) means that aspect of reality that is not accessible to the five senses, but may still be understood by rational thought. We noted earlier that another term for level (1), the world of appearance, is the empirical realm. Realm (2), the underlying or hidden reality, is also what Kant called the noumenal realm.

[1]This chapter is loosely based on a portion of Kastner (2012, Chapter 8). ©2012 Ruth E. Kastner. Reprinted with permission.

We also noted in Chapter 6 that even though physics is an empirical science, by which we mean that its predictions must answer to empirical data, physics need not be restricted in its subject matter only to observable entities. That point was demonstrated by the success of Ludwig Boltzmann's pioneering approach (summarized in Boltzmann, 1896). Recall that Boltzmann was able to obtain the laws of thermodynamics governing the behavior of observable macroscopic objects by devising a theory about the behavior of underlying, unobservable microscopic objects; i.e., atoms. Moreover, his pioneering approach led to crucial breakthroughs in physics. So we can correctly say that the task of physics is to attempt to describe all of reality, including aspects of it that may not be observable to us, even in principle.

The way that physics describes reality is by insightfully analyzing the data gleaned in an empirical investigation, using logic and mathematics. This analysis may include creative hypotheses concerning unobserved or even unobservable objects. That's where the insight and imagination enters; the kind that Boltzmann used when he hypothesized the existence of atoms. Einstein, probably the most famous physicist to expand on Boltzmann's approach, said: 'Imagination is more important than knowledge. For knowledge is limited, whereas imagination embraces the entire world...' (Einstein 1931). Thus, physics imaginatively, yet also rationally, studies the empirical realm in order to understand both the empirical and sub-empirical (unobservable) realms.

Of course, a strict empiricist would deny that the job of physics is to gain knowledge of a sub-empirical realm, even if it exists, since an empiricist denies that anyone can gain knowledge of anything not based on observation.[2] But recall that when Mach strongly objected to the existence of atoms, on the empiricist basis that they were unobservable, he turned out to be on the wrong side of scientific progress. Based on this history, we might reasonably conclude that the strict empiricist approach is not a

[2]The empiricist agrees that we can have knowledge about the truth of necessarily true kinds of statements, called 'analytic' statements in philosophy. An example is 'All bachelors are unmarried.' In contrast, a knowledge claim that is not necessarily true by its own structure is called 'contingent'. The empiricist asserts that the only way one can gain contingent knowledge is by observation, so an empiricist would deny that one can gain contingent knowledge about an aspect of reality that is not observable in principle.

fruitful one, and that it is indeed scientifically valid to consider the existence of unobservable entities that may be of explanatory value in understanding the inner workings of the phenomena that we *are* able to observe. In this chapter, we'll consider in more specific terms the idea that there is indeed an unobservable aspect to reality, one which transcends spacetime.[3] To that end, in the next section we'll examine the spacetime construct in more detail, and see that it is a limited one that need not constrain our thinking.

The Spacetime Map

Suppose you are embarking on a road trip to the beautiful Adirondack Park in upstate New York. (New York is a huge state containing expansive rural and mountainous regions, and the Adirondack Park encompasses much of that area.) Your friend has invited you to his cabin near Schroon Lake, and since you are unfamiliar with the area, you have brought along a good map of the park. As you head up Interstate 87, you begin to see signs that the park entrance is approaching, so you get out your map to help you find your way around. Now, suppose I were to draw a chart representing the positions of your car at particular places and times (Figure 7.1).

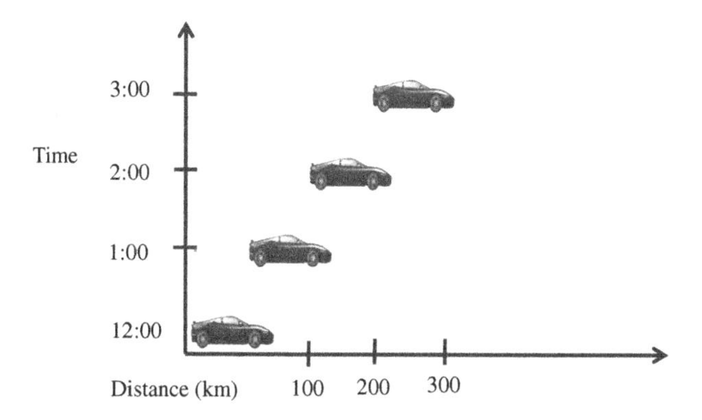

Figure 7.1. A spacetime map of your car's trajectory along Interstate 87.

[3]Interestingly, Bohr himself once remarked that quantum processes 'transcend the frame of space and time' (as quoted in Petersen (1963)).

This 'spacetime diagram' illustrates the situation at various times after your departure (say at 12:00 p.m.), heading for your friend's cabin in the Adirondack Park. In the diagram, we have three spatial dimensions shrunk down to only one horizontal line. You can think of that horizontal line as the road you're traveling on inside the park. Meanwhile, the vertical line represents the passage of time as would be indicated on a nearby clock. Each tick mark on the vertical line indicates the passage of one hour, and each tick mark on the horizontal line represents about 100 kilometers. If you like, you can think of this as a series of frames in a movie clip, or like the pages of a flip book. The car advances as you flip the book; in the same way, the car advances as you move your gaze upward on the diagram in the vertical direction representing the passage of time.

Thus, we can represent some aspects of your experience on this 'spacetime diagram.' But it is only a map that illustrates certain aspects of your experience, and it can lead to unsupported assumptions about what reality is like. For example, suppose you look at your watch and find that it reads 3:00 p.m., meaning that at this moment, you have reached the 300-kilometer mark. You might be tempted at this point to think that you could extend the diagram into the future by drawing something like Figure 7.2.

Remember that this diagram represents your past experience and your present experience as your watch reads 3:00 p.m. The dotted box around

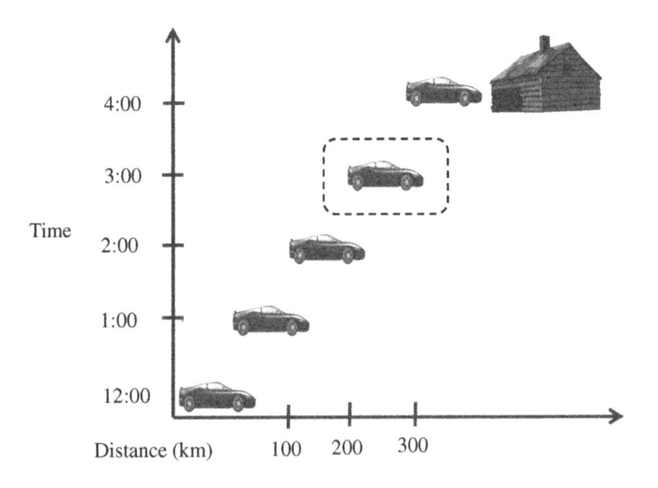

Figure 7.2. A spacetime map with a 'future event.'

your car at 3:00 p.m. serves to indicate that this is your present experience, at the 300-kilometer mark. Therefore, this diagram illustrates the following situation: in the future, at 4:00 p.m., you will arrive at your friend's cabin, and that event will definitely occur and cannot be avoided (it is your 'fate'). Your arrival at the cabin is shown in your future because it is displaced in the vertical direction, which represents time.

The fact that we can draw pictures like this — that is, we can arbitrarily place events in the future with respect to any other event on a spacetime map — has tempted many researchers to conclude that there really are 'future events' that exist in spacetime in exactly the same way as present and past events. For example, it is very common practice among physicists to identify some point on a spacetime diagram as 'Now,' and then talk about a 'future observer' who is placed, like our log cabin, vertically above the point identified as 'Now'. Along with this idea of already-existing future events often goes the assumption that there is some real, substantial entity called 'spacetime' that acts as a container for all the past, present, and future events. This kind of picture is called a 'block world,' and many physicists subscribe to it.

Besides the ability to insert 'future' events into the spacetime map without restriction, favorable views of the block world idea are based on certain aspects of relativity theory. One of Einstein's contemporaries, Hermann Minkowski, put it this way: 'Henceforth space by itself, and time by itself, are doomed to fade away into mere shadows, and only a kind of union of the two will preserve an independent reality' (quoted in Newman, 1956). Minkowski came to this belief because of the way that relativity theory deals with the differing spatial and temporal measurements of observers in different states of motion (i.e., observers moving at different speeds relative to each other). The equations needed to express the relationships between different observers' spatial and temporal coordinates involves a kind of intermingling of those coordinates. Minkowski showed that these relationships, called 'Lorentz transformations,' could be expressed on a spacetime map by tilting the spatial and temporal axes towards each other (Figure 7.3).

In Figure 7.3, the tilted gray axes are those of an observer moving relative to the one whose axes are perpendicular. The faster an observer is going relative to another observer at rest, the more tilted his spatial and

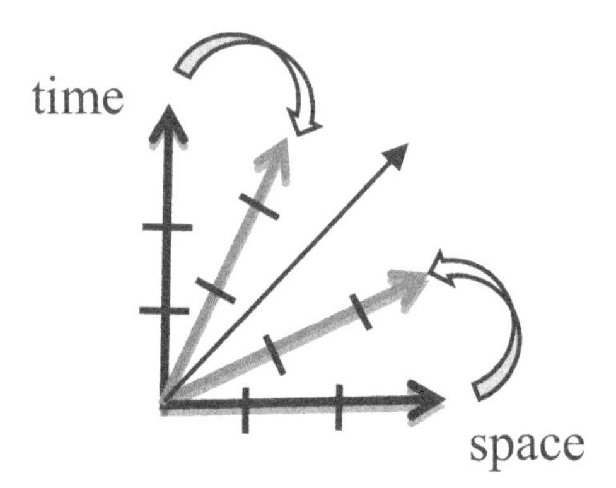

Figure 7.3. A moving observer's spatial and temporal axes are tilted towards each other. The diagonal arrow represents the speed of light.

temporal axes. The tilted temporal axis represents the speed of the moving observer, just as the car traces out a tilted line in Figures 7.1 and 7.2. The surprise presented to us by relativity theory is that the spatial axis also tilts up from its horizontal position; this is noticeable only for very fast-moving objects. Another surprise is that the tick marks are stretched out on the tilted axes; these are the time dilation and length contraction effects. (We didn't worry about these in the case of the car, because such effects are only significant at speeds approaching the speed of light.)

The fastest possible speed in spacetime is the speed of light, and this is indicated on the spacetime diagram in Figure 7.3 by the thin diagonal arrow.

The Map is not the Territory

As described in the previous section, one can certainly express relativity theory in terms of a unified spacetime called 'Minkowski space.' However, this representation is a kind of 'map' that captures certain crucial aspects of relativity theory while smuggling in other features that may not be crucial at all. One such unnecessary feature is the idea that there is a substantive spacetime container sitting there, like a piece of graph paper, occupied by various past and future events. Another unnecessary feature is the idea that all spacetime events — past, present, and future — exist, and that there is no fundamental difference between such types of events.[4] This chapter will argue that this is not the right way to think about space and time, and will propose a different understanding.[5]

First, it's important to note the distinction between the map and the territory. The map is the spacetime representation, but the territory is reality, which includes observable events as well as unobservable supporting entities such as quantum systems; those cannot be shown on the map. And, just as there are aspects of reality that cannot be shown on the map, there are aspects of the map that don't necessarily correspond to true statements about reality. (The 'block world' aspect of the map is one of these.) For example, just because the Adirondack Park map is drawn on a piece of paper does not mean that there is a substance in the real world corresponding to the paper that the map is drawn on (i.e., a spacetime substance or 'container'). And an event drawn on a spacetime diagram in someone's future does not necessarily represent anything that exists in the real world either, nor does it prove that future events exist in the same way as present and past events. Even if you can draw your arrival at your Adirondack cabin on the spacetime diagram, it does not mean you will necessarily get there. Your car could break down or run out of gas.

Thus, the spacetime diagram, because it is so easily subject to arbitrary event placements in a hypothetical 'future,' typically misleads us into thinking that there can be 'future events' and 'future observers' when this is physically not the case. Just because we can draw something on a spacetime diagram does not mean that it can physically exist in our world. The idea that our ability to draw something on a spacetime diagram implies that it may physically exist can be very compelling. To see why we need to be wary of this, here's an interesting analogy from the craft of

---

[4]The block world is a view known as the 'B series of time' in philosophical discussions of time. The alternative view, that there are important differences between past, present, and future events, is called the 'A series.' This terminology comes from a famous discussion of the nature of time by McTaggart (1908).

[5]I am not the first to question the idea that relativity requires a substantive spacetime 'container.' Harvey Brown (2002) extensively critiques that view. For additional arguments about why relativity need not imply a block world, see Kastner (2012, Chapter 8) and Tooley (1997).

animation. Animation artists now have programs that can do a lot of the work of redrawing many frames of the same character for them. A typical animation program allows you to load an image of a character and to indicate where all the joints are. Then the program will incrementally change the angles of the joints for you, in a series of images that can make the character appear to move. All you have to do is to specify the amounts that each of the joints should move in each frame. This kind of program may allow you to make a character's head turn by any amount in any direction, but that doesn't mean that such a motion will be realistic or even physically possible.

In much the same way, the ability to draw any event wherever we choose on a spacetime diagram does not imply that what we drew corresponds to what is physically possible, any more than the ability to make a character's head spin around in circles in an animation program demonstrates that this kind of motion would be possible in the real world. To explore these issues further, we'll need to take a careful look at what is meant by the 'empirical realm' — that is, the world of appearance — which is so important to physics, yet which can still be deceiving. The construction of representations of the empirical realm can lead us to smuggle in notions (such as a spacetime 'container') that may not really apply to reality; and by the same token, to neglect others that do.

What is the Empirical Realm?

What is the empirical realm, the world of appearance, in physical terms? Physicists generally think of all of spacetime as the empirical realm. This is because they can represent observable events on spacetime diagrams such as the one in the previous section. However, this can't really be right if the empirical realm is the realm of direct experience. This is because we never actually experience either the past or the future, even with powerful instruments. It takes light from the Sun about eight minutes to get to the Earth; that distance is 93 million miles. If the light we get from the Sun left its source eight minutes ago, we might be tempted to think that we are 'looking into the past' when we see the Sun in the sky. However, we don't actually see that light until it reaches us in the present. So we see the Sun in the present as it was eight minutes ago. Similarly, when we see a galaxy that is two million light-years away, we are seeing it in the present as it was two million years ago. This is essentially the same as getting a message in a bottle from a castaway. The sender of the message may be long dead, but the message is something he wrote while alive. So we don't actually experience the past. All we ever directly experience is the present moment, the Now, as it is presented to our senses.

The same can be said on a much smaller scale for another person in the same room as you. It takes light about three billionths of a second to travel one meter. If another person is standing three meters from you, that person is seeing you not as you are, but as you were about 10 billionths of a second ago. That also goes for everything you both see in the room. If one of you is closer to, say, a very precise clock, that person will see a very slightly different time than the other person, who is farther away, (although by a just a few billionths of a second, an imperceptible amount by human standards.)

So, if we want to be careful about it, only the Now is the empirical realm. What's the Now actually doing? Is it in motion? No. Although we commonly think of the Now as 'moving forward in time,' this is not what we actually experience. What we actually experience about Now is that it exhibits properties to us that are always changing. As John Norton (2010) observed, 'we do have a direct perception of the changing of the present moment. That is clearest in our perception of motion.' In other words, the Now itself does not really 'move,' it changes.

How do we experience these changing properties of the present moment, or Now? We experience them through electromagnetic signals that transfer energy from emitters to our sense organs. As discussed in Chapter 3, these transfers of energy are brought about by way of actualized transactions. So any given property of our 'Now' is defined through touch by transactions between ourselves and the object(s) with which we are currently in direct contact, and through sight by transactions actualizing the photons that are reaching our eyes from other objects. This is illustrated in Figure 7.4, in which the Now is symbolized by the photo of the person (let's call him Chuck), and the light signals reaching his eyes from objects in his past, such as the two stars depicted on either side below him. Past events can also make their presence known to Chuck by way of slower signals, such as a basketball pass (this is indicated by the line coming from the man shown in the central part of the past light cone).

Figure 7.4. The 'Now' is the empirical realm.

There is another aspect of the spacetime diagram that can be misleading, and that is the notion that we can extend our Now beyond ourselves into our surroundings, and onto other events. This is one of the notions alluded to above that is 'smuggled in' via the map, but does not necessarily apply to the territory. Here's how it's typically done in such diagrams: we draw a horizontal line extending out from our location and as far into space as we like. This so-called 'line of simultaneity' is labeled by whatever time appears on our watch; for example, 12:00 p.m., as in the charts of Figures 7.1 and 7.2. We can draw as many horizontal 'lines of simultaneity' as we wish, each at a different time interval, such as those indicated on the vertical axes of Figures 7.1 and 7.2. But the fact that we can assign '12:00 p.m.' to other points in space on a diagram does not mean either that other 'points in space' really exist or (even if they did) that they would necessarily agree that it is 12:00 p.m. For one thing, there is no way we could ever communicate with any entity at these locations, since that would require a signal faster than light.

The basic point is this: since electromagnetic signals have a finite speed, no observer really sees anything the way it exists at the same instant as it is perceived. When you sit in a chair reading this book, you are seeing the page as it existed a few nanoseconds ago (because that is how long it took the light to reach your eyes from the page), not as it exists along a line of simultaneity from your eyes.

So the Now is an individual, local phenomenon, and it constitutes each individual's empirical realm. We often think of the empirical realm as objective in the sense that it contains phenomena that can be corroborated among many different observers. However, the empirical realm can't be truly objective in an absolute sense, because it's defined in terms of appearance, and appearance can only be relative to a given observer. That is, each person will experience a different realm of appearance, just as in Bertrand Russell's observations about the table, as discussed in Chapter 5. This means that, strictly speaking, every individual has his or her own private empirical realm. That's what makes it so hard to define the 'real,' objective table, or any other perceived object.

However, we can corroborate our experiences and arrive at a consistent public consensus about a 'larger' world of appearance beyond our individual empirical realms. All of these corroborations are conducted using electromagnetic waves. This tells us that 'the empirical realm' in physics is not really all of spacetime, as is often assumed, but instead is a well-coordinated collection of individual Nows. Another way to characterize the empirical realm is as a unified set of events that describe a coherent whole. The whole is coherent because the events making it up fit together according to regularities, or laws, that describe the behavior of events in ways that can be corroborated. These are the laws of physics.

The Fabric of Spacetime

In the previous section, it was argued that future events don't exist in the same way as present and past events, despite our ability to put a 'future event' anywhere we wish on a spacetime diagram. In this section, we're going to look in more detail at how the spacetime realm arises from the quantum level of possibilities in the transactional picture. This process is strikingly similar to the process of knitting a piece of fabric using yarn and needles (see Figure 7.5). When we knit, new yarn from the ball is pulled through the stitch currently on one needle with the other needle, to make a loop; that loop becomes a new stitch on the needle that formed it. Meanwhile, the stitch that was there before drops off the needle and is extruded as a stitch in the fabric. Using this analogy, we can immediately see what distinguishes the Now from all other aspects of reality: the Now, or present, for any given observer, is the set of stitches on the knitting needles.

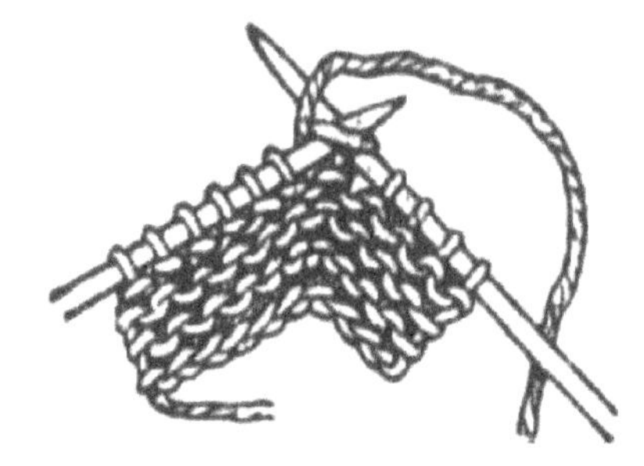

Figure 7.5. The past as a knitted fabric; 'Now' corresponds to the set of stitches on the needles.

Naturally, as the process of knitting continues, the stitches on the needles change. But, just as the needles don't move anywhere, the Now does not move anywhere (or 'anywhen'). Instead, it is the fabric that is extruded; the stitches in the extruded fabric are the events of the past. In our analogy, that fabric of events is what we think of as 'spacetime.'

Let's consider what happens at the very beginning of the knitting process: the 'Big Bang,' if you will. In knitting terms, this is where the initial set of stitches (think of these as the primordial cosmic particles) is cast on to the empty needle. This creates our first 'row' of stitches, which we can index with numbers representing times. So, for example, the first row can be labeled as $t = 1$, the second row as $t = 2$, etc.[6]

In the transactional picture, what creates a new stitch in the spacetime fabric is the actualization of a specific transaction. Recall from Chapter 5 that not only the transmitted quantum but also the participating emitter and absorber become localized as spacetime objects due to an actualized transaction. Suppose now that the absorber is a cell in your eye. The absorption event of the actualized transaction defines the Now, and the emission event has already been extruded at this point. So that means that the emission event of an actualized transaction is always in the past relative to whomever

[6]Readers familiar with relativity may object that each row seems to play the part of a 'line of simultaneity,' which was just rejected as unphysical. This is a limitation of the analogy, which cannot faithfully represent the creation of events having a 3 + 1 dimensional Minkowski space structure. See Kastner (2012), Chapter 8, for why a block world is not forced upon us even though this particular analogy has this shortcoming.

is experiencing that absorption event. In this sense, the actualization of any transaction necessarily pushes an event 'into the past,' and that event is the emission for that transaction. This is how spacetime is created: events are extruded away from the Now, and into the past. We'll pick up on this point below, when we take a look at the 'delayed choice experiment.'

In addition, it should be noted that 'Now' is defined by any absorption event, and only with respect to that specific event: it is a highly localized concept. Significantly, the French word for 'now' is '*maintenant*'; literally, 'holding in the hand.' What do you 'hold in your hand' but the energy received from all the emitting objects around you? This emphasizes the inescapably local quality of Now and the idea that it is the domain in which events are created through the transfer of energy between emitters and absorbers. We directly experience our sense of Now through the absorbers in our sense organs and in our brains.

The quantum possibilities are represented by yarn of various colors and types, as well as the various patterns and/or ideas about what to knit and how to knit it. So, in this picture, just as there are no stitches in fabric that haven't been knitted yet, there are no future events: the future is just those possibilities. The Now is the realm in which our garment is created; the Now doesn't 'move,' but the stitches on the needles change (perhaps in color or texture) and are extruded away from us in the form of fabric as the knitting progresses. Thus, the Now is not something that 'moves forward'; rather, the Now is the empirically-always-present field of change, while the past is something that continually falls away from us. Meanwhile, the future is a collection of dynamic possibilities which exist as real physical entities, but which may or may not become part of the spacetime fabric.

Delayed Choice Experiments: Playing Games with Spacetime

Now that we have a basic model of spacetime in the transactional picture, we can begin to make sense of some other perplexing quantum riddles. In this section, we'll consider a class of experiments based on an idea by John Wheeler. In 1978, he proposed a now-famous thought experiment that highlights the strange indeterminacy of quantum objects.[7] (The experiment

[7]The thought experiment was presented in Marlow, A. W. (ed), 1978.

has since been performed many times.) It's a variation on the two-slit experiment. In what follows, I'll use the usual sort of language to describe the experiment, but the reader should try to keep in mind that this language is fundamentally misleading. It's misleading because it describes quantum objects as if they are following spacetime trajectories, but they are not really doing so. We'll also see that when we describe what's really going on in the transactional picture, we have a more illuminating account, even though it's not a spacetime account.

Now, recall that (in the usual sort of distorting language) if we arrange to detect which slit the photon went through, we lose the interference pattern and cause the photon to go through only one slit or the other. That kind of measurement can be made by focusing telescopes on each slit. On the other hand, if we don't arrange to detect which slit the photon went through, the photon goes through both slits and we see interference. Wheeler proposed the following twist: what if we don't decide which kind of measurement to make until after the photon has already gone through the slits? We can do this if we have a removable screen to detect the interference pattern, while the option of measuring which slit the photon went through is performed by telescopes focused on each slit, as discussed in Chapter 3. In order to choose the 'which-slit' measurement, we quickly remove the detection screen after the photon has gone past the slits, but before it hits the detection screen. The setup is shown in Figure 7.6.

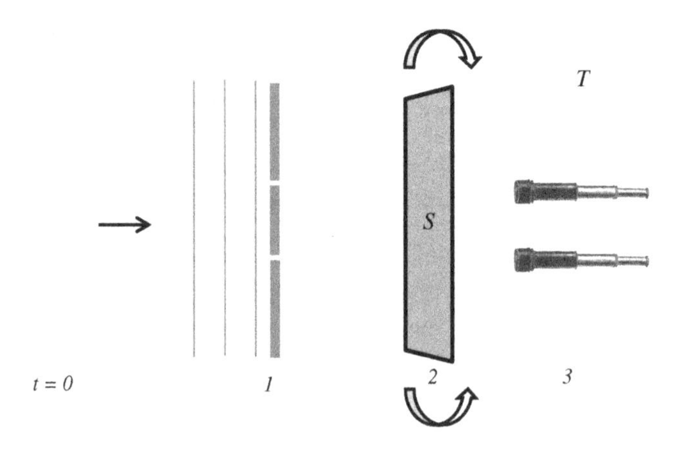

Figure 7.6. The delayed choice experiment.

In Figure 7.6, a source, such as a laser, emits a photon quantum in the direction of a screen with two slits. The thin vertical lines approaching the first screen represent wavefronts (remember that the photon's quantum state is based on the de Broglie wave, and in the transactional picture it's an offer wave). The number of seconds elapsed after emission of the photon are shown at the bottom. (Actually, light travels so fast that it is not really 'seconds' but much faster than that. But using seconds as our unit of measure helps us to visualize the experiment in time frames we are familiar with.) Behind the screen with the slits is a removable detection screen, *S*, and behind that are two telescopes, one focused on each slit. The screen with the slits is placed where the photon is expected to be one second after it is emitted, the detection screen, *S*, where the photon is expected to be two seconds after it is emitted, and the telescopes where it's expected to be three seconds after it's emitted.

When screen *S* is in place, we get the usual interference pattern on that screen, since the photon never reaches the telescopes; this is a 'both slits' measurement. On the other hand, when screen *S* is not in place, the top telescope detects only photons that go through the top slit, while the bottom telescope detects only photons that go through the lower slit. So this would be a 'which-slit' measurement, yielding no interference. Suppose we release a photon quantum from the laser, and leave screen *S* in place. For a single photon, we will detect just one dot on the screen, but if we continue to let photons accumulate with the screen in place, we'll eventually see an interference pattern of bright and dark stripes, as we saw in Chapter 2 (Figure 2.7). In the usual way of talking, this means that each of those photons went through both slits. On the other hand, for each of these photons, we could, if we wish, remove screen *S* after the photon passed the screen with slits, but before each of those photons got to screen *S*. When we do this, they are detected at one or the other telescope, with no interference occurring. That is clearly a 'which-slit' detection, which means (in the usual way of talking) that each photon would have to have gone through only one slit or the other. But each photon had already passed the screen with the slits, and did whatever it did, *before* we made that choice to remove screen *S*. So we have an apparent contradiction: in the usual way of thinking, the photon went through both slits while screen *S* was in place, but after we removed screen *S* it appears as though the photon only went through one slit. The delayed choice experiment has been performed in the lab, and indeed, whether or not you see an interference pattern does depend on what kind of measurement you choose to make; i.e., whether you choose to remove the screen or not.

Thus, the bizarre feature of this experiment is that it suggests that our choice of which kind of measurement to make between one and two seconds after the photon has been emitted has effects that seemingly reach into the photon's past, to an earlier time. If we let the interference occur by choosing to leave screen *S* in place, then the photon went through both slits one second after it was emitted. On the other hand, if we do a 'which slit' measurement by removing the detection screen, the photon only went through one or the other slit one second after it was emitted. The photon did whatever it did before we made our choice, so apparently we must be influencing the photon's past behavior by making our choice!

How can we understand this particular quantum riddle with the help of the transactional picture? Recall that in transactional interpretation (TI), what is emitted is not a particle that follows a spacetime trajectory, so we shouldn't be envisioning a photon as a little particle zooming through space on a specific trajectory, as the above sort of description implies. Instead, what is emitted is a quantum possibility: the offer wave. This is not a spacetime object; it lives in 'Quantumland.' Yet, it can interact with our lab equipment, which itself is made up of quantum possibilities at a fundamental level.

To understand the TI picture of the two-slit experiment, consider the apparatus depicted above, in which only one photon at a time is emitted. To keep track of things, we'll refer to some specific times (the units don't matter, but you can think of these as seconds.) What the source emits at $t = 0$ is not a little particle traveling through space. Instead, it emits an offer wave with a certain momentum and energy.[8] The offer wave continues through both slits to whatever absorbers are available to it at $t = 2$ *or* $t = 3$, either the detection screen or the telescopes. Then, whatever absorbers are stimulated by the offer wave generate confirmations. It is those confirmations that determine what kinds of incipient transactions will be set up, and thereby determines the resulting detection phenomena.

[8]Here we're disregarding runs of the experiment in which the photon gets absorbed by the first screen with the slits.

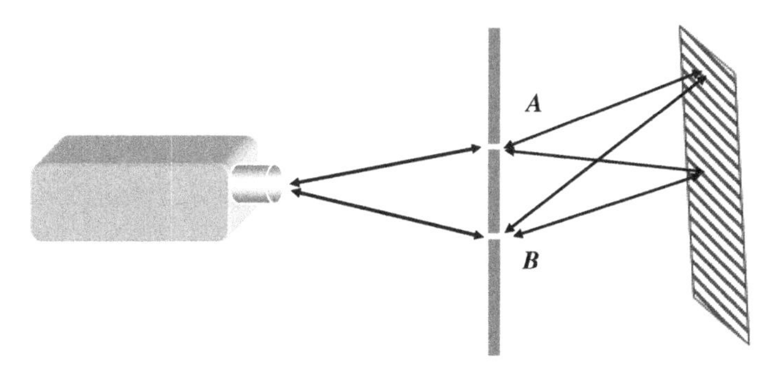

Figure 7.7. The two-slit experiment with interference detected. The offer-wave components reaching each absorber in the screen come from both slits, and their confirmations also go through both slits.

If the offer wave meets the detection screen, each absorber in the screen receives a different component of the offer wave and responds with a confirmation matching that component (see Figure 7.7). In Chapter 3, we considered just two sample absorbers in the screen, but the entire screen is composed of tiny absorbers. Each of the absorbers responds with a confirmation matching the component of the offer wave that reached it. All components of the offer wave reaching each absorber went through (interacted with) both slits, so all the confirmations also interact with both slits, since the confirmations are mirror images of the parts of the offer wave that reached them. This sets up a set of incipient transactions, all of which are 'both slits'-type transactions. Only one of these is actualized for each photon absorbed at the screen, but because the set of incipient transactions are 'both-slit' types, an interference pattern is built up when many photons are detected, as discussed in Chapter 3. In this case, the absorption events are actualized at $t = 2$, relative to our laboratory clock.

Now, consider the case when the screen is removed so that the offer wave can reach the telescopes (see Figure 7.8). The emitted offer wave itself is unaffected, as it still goes through both slits, just as before. But in this case, the individual offer wave components that reach each of the two telescope absorbers are different from the ones that would have reached each of the many absorbers in the screen. To see in details why this is, we would have to go into the details of how telescopes work. But, in a

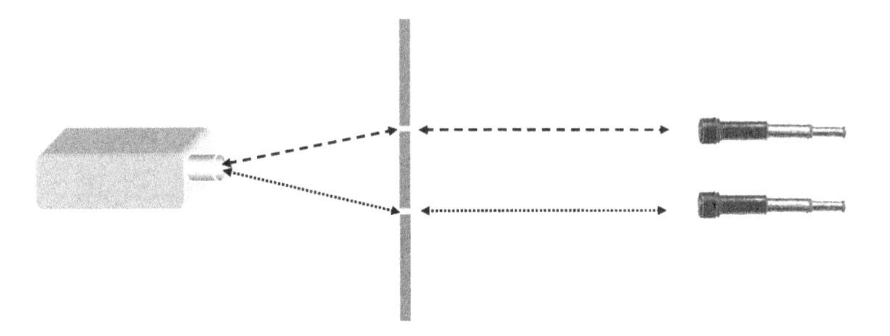

Figure 7.8. The two-slit experiment with a 'which slit' measurement. Only a single-slit offer wave component can be absorbed by each telescope, and each therefore returns a single-slit confirmation. The A-slit components are shown as a dashed line, and the B-slit components are shown as a dotted line.

nutshell, each telescope can only pick up the offer wave component that carries a signature of which slit it went through; specifically, it must have a very precise directional momentum that can only correspond to passage through one slit or the other.

The crucial point is this: the offer wave still goes through both slits. But the part of the offer wave that has interacted with the upper slit can only reach the top telescope, and the part of the offer wave that has interacted with the lower slit now can only reach the bottom telescope. Each telescope responds with a confirmation wave that matches only the corresponding part of the offer wave that reaches it, and the confirmation goes back through that same slit. In this case there are two distinct incipient transactions, each corresponding to the part of the offer wave that interacted with one or the other slit and reached the corresponding telescope.[9] That's why this is a 'which slit' measurement as opposed to a both-slits measurement. In this case, the absorption events are actualized at $t = 3$, relative to our laboratory clock.

So, in the transactional picture, what we see is that the offer wave is unaffected by our delayed choice, at least in the sense that it always interacts

[9]The imaging portion of each telescope is actually composed of many microscopic absorbers. So there may be many competing incipient transactions, just as many as with screen S in place, but at the macroscopic level we'll only be able to distinguish these two possible results.

with both slits.[10] It is only the confirmation that is affected by that choice. Our influence on the past is limited to the confirmation wave that is generated, because, as noted above, that is how the past is created. At the quantum level, the past is always dependent, to at least a slight degree, on what happens in the present, just as the knitted fabric is extruded from the needle in a way that depends on the knitting action and the yarn. The generation of a confirmation in the present is what dictates whether, and by what sort of process, energy is removed from an emitter that must become part of the past when the transaction is actualized. This is because, for any actualized transaction, the emitter must always be actualized in the past relative to the absorber. It's important to remember that the transferred quantum begins the transactional process outside spacetime, in 'Quantumland.' It only becomes part of spacetime upon the actualization of the transaction. Then the emitter takes its place in spacetime as part of the past, while the actualized, transferred quantum connects the emitter to the absorber in the present.

There is a nice way to visualize this process in terms of the knitting metaphor. First, remember that quantum processes underlie all of the phenomena we see, including those that appear 'classical.' This is because all phenomena are the result of actualized transactions. The classical realm is the scale at which we can get away without having to keep track of all the underlying offers and confirmations and transactions. Remember now that the stitches in our knitted fabric represent actualized transactions, so think of the fixed, ordinary, classical parts of the experiment as a knitted spacetime fabric made out of very small, tight stitches. That is, you have to get a magnifying glass to be able to see that they are stitches (i.e., quantum transactions) at all, so that if you wanted to, you could pretend that your fabric was completely smooth and solid. But suppose we wish to work specifically with quantum objects: this corresponds to a loosely-knit fabric with big, discernible stitches. We can get even fancier with the delayed choice experiment. Here's how we do it: it's a procedure very much like knitting a 'cable pattern.'

---

[10]Although you might be tempted to think that 'interacted with the slits' means the offer wave had to have been in spacetime, this is not the case. The components of the slits, at the microscopic level, are quanta themselves. The interaction of the offer wave with the slits takes place in Quantumland, *beneath* the 'tip of the iceberg.'

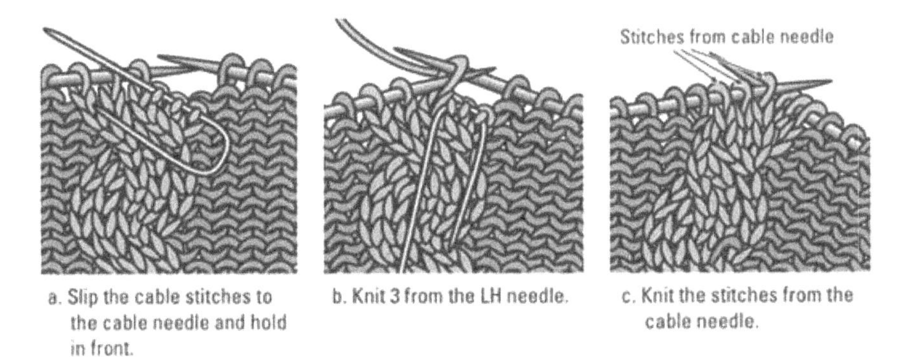

Figure 7.9. A cable stitch is like a quantum delayed choice experiment.[11]

In Figure 7.9(a), some stitches are removed from the knitting process and held in 'standby' on the cable needle (the U-shaped needle). In Figure 7.9(b), the surrounding stitches are knitted into the 'past'; this represents the ordinary, classical, experimental equipment. In Figure 7.9(c), the 'delayed choice' is made and the stitches in standby are taken up and knitted back into the fabric. The result is a pattern with more texture and depth than the plain 'classical' fabric.

The basic point is that the offer wave can remain as an indeterminate, quantum entity, even in the past from our perspective, until a confirmation is generated. Its indeterminacy is represented in this analogy by the way the stitches are held in a separate 'Now,' on the cable needle, and are not yet incorporated into the rest of the garment. The confirmation does not affect anything in the past that is already part of spacetime. But the offer wave is not contained in spacetime; it is still in Quantumland, on 'standby.' This is represented in the analogy by the cable stitches being slipped onto the cable needle and 'held in front'; they are literally not part of the spacetime fabric during this process. So even though the generation of the offer wave can be given a time index (in this case the row number), that does not mean that the offer wave itself is restricted to spacetime. The offer wave will give rise to a spacetime event (a new stitch in the fabric)

[11]Image from Allen, P., Barr, T., and Okey, S., *Knitting for Dummies*, 2nd Ed., Wiley Publishing, Inc. © 2008 by Wiley Publishing Inc., Indianapolis, IN. Reprinted with permission.

only if it is confirmed and the resulting incipient transaction is actualized. In the metaphor, we can (in rough terms) think of the confirmation as the stitch being picked back up on the needle, and the actualization as the new length of yarn being knitted into the stitch. While the knitting action (confirmations and actualized transactions) on the cable stitches is suspended, those stitches are being held on a needle, in a kind of Now of their own. The cable needle represents this auxiliary 'Now.'

Just as in the knitting process, the creation of events is a process of stitching between the past (i.e., the stitches being extruded) and the future (the balls of yarn providing the raw materials). In philosophical terms, this is a kind of 'growing universe' theory of time. But unlike most such theories, which envision the Now advancing into the future from a fixed starting point, in this picture it is the past that grows and continues to become actualized as it falls away from the present. The 'Now' is what is fixed, because the Now is the eternal field of creation of the spacetime fabric. Meanwhile, the future is not a realm of determinate events, but rather a realm of physical possibilities; it is the 'raw material' for events. The future is a set of possibilities that becomes woven into the created past through the action of Now.

Spacetime is just Actualized Events and their Relationships

In this chapter, we've seen how the transactional process 'weaves' the fabric of spacetime through actualized events. It is the events themselves that constitute spacetime; there is no spacetime substance apart from those specific events, nor any empty spacetime 'container' waiting to hold the events. If this seems hard to visualize, consider a circle formed by children playing (Figure 7.10)

There was no circle before the children arrived and joined hands. The circle represents a spacetime structure, and the children represent events making up that structure. We see that it is the events that create that structure we call spacetime, and without those events, there is no spacetime. Thus, the transactional picture provides an elegant account of the emergence of the set of spacetime events from the quantum level. There is no substantive spacetime 'container,' there are simply actualized events. This picture provides a new way of understanding why our sense of 'Now' is

Figure 7.10. The painting 'Children's Dances,' by Hans Thoma.

so inescapable, yet is difficult to find in classical spacetime theories. The Now is our empirical realm, in which quantum transactions are actualized, and the events that they actualize fall away from us as the past. The future is 'nothing but possibility.' But that is real quantum potentiality, and it is an essential ingredient in the creation of the event-based fabric of spacetime.

In the next chapter, we'll look more deeply into the nature of time.

Chapter 8

Time's Arrow and Free Will

'Yesterday is history, tomorrow is a mystery.'

Eleanor Roosevelt

In the previous chapter, we saw how the 'fabric of spacetime' is created through actualized transactions. In this chapter, we'll study more closely the nature of that spacetime fabric, and see how it allows for an 'arrow of time,' pointing in the future direction. We'll see how that future is not set in stone, but is genuinely open to many different possible events. This openness of the future provides an opening for genuine free will, rather than the necessary predestination of the block world picture.

But first, consider lightning.

Figure 8.1. Lightning starts from one region in the cloud and keeps its options open as to where to strike.

Lightning, as pictured in Figure 8.1, consists of a flow of charged particles from a cloud toward the ground. But notice that the flow begins at a specific point in the cloud and often branches, striking several different points on the ground (or at least aiming for them). The reason we're considering this is because it illustrates a key feature that underlies the asymmetrical orientation of time toward only one direction, which we call the 'future.' The asymmetry results from the fundamental difference between an emitter and an absorber. The emitter is the starting point of any transaction. In addition, there are (in general) many possible absorbers for one emitted offer wave. This asymmetry between the emitter and its responding absorber(s) is the reason, in a nutshell, for the apparent flow of time that we experience. In the following, we'll crack open that nutshell and see how it all works.

Physics and Time

Physical laws are often taken as indicating that time could flow in either direction and therefore that the apparently unidirectional flow of time is an illusion. This is because the laws of motion seem to be reversible with respect to time: they can be just as easily run 'backward in time' as they can 'forward in time.' For example, of you were to look at a film clip of magnified gas molecules in motion, you would not be able to tell whether it's being run in the forward or backward time direction. Both would look like realistic physical processes.

Thus, the basic laws of motion (both classical and quantum) don't seem to support the idea that there is a unidirectional flow of time. If you add to this the arguments we considered in the previous chapter for a 'block world' (in which there is no fundamental distinction between past, present, and future), you can see why many physicists want to conclude that, as Ford Prefect says in Douglas Adams' *Hitchhiker's Guide to the Galaxy*, 'Time is an illusion; lunchtime doubly so.' In a certain sense, Ford is right: time isn't something that literally exists as a concrete substance that 'flows.' Rather, time is the measure of change. But change is very real; and despite those time-symmetric laws, it is change that gives time its directionality. The directionality of change comes from the transactional process, which introduces an asymmetry in the otherwise symmetrical laws. In this chapter, we'll see how this works.

What is the root of change? It is energy transfer from one thing to another. 'A time interval' is only defined relative to change, and change is defined by differences between events. Therefore, a time interval is only

Figure 8.2. A time interval is just a comparison of two different events.

meaningful in terms of reference to two specific events. For example, a clock (Figure 8.2) is a system that changes its state in a predictable way, and we measure time intervals with a clock by comparing two different states of the clock.

Each observation of those two different states is an event. So we are never really 'measuring time'; we are just comparing two different events in space. That's all a clock really does. It's tempting to think, as did Isaac Newton, that time passes independently of these sorts of physical changes, but there are good reasons to reject that view. A major reason is that relativity theory tells us that a time interval is only defined relative to a particular observer; it is not absolute. Two events that appear to be simultaneous as measured by one observer are not simultaneous as measured by another who is in motion relative to the first observer.

Before we consider in more detail the relationship of energy transfer to change: as I noted above, not every single physical law is time reversible. One physical law that is not time reversible is the collapse process described by the Born Rule, which we have discussed previously. Remember that the Born Rule gives us the probabilities of outcomes of measurements performed on quantum systems. It is not reversible because it describes a collapse, in which an initial quantum state indeterministically changes to one of several options. Once collapse has happened, information has been irretrievably lost, and you can't 'rewind the film clip' and get it back. But because the physical basis of the Born Rule is not understood outside the transactional picture, it has generally been excluded from consideration as a fully-fledged physical law. In contrast, I've argued in this book that the Born Rule describes a real, time-asymmetric physical process: the actualization of one of a set of incipient transactions. The actualized transaction irreversibly transfers energy and other physical quantities from its source, the emitter, to the receiving absorber. In doing so, it changes both of them.

Energy and time

One of the curious facts about quantum theory is that certain pairs of observables have special relationships. We explored this in Chapter 2 when we saw that some observables are incompatible. The prime example is the incompatibility of position and momentum: measurements of position and momentum disturb each other. But this incompatibility is also evidence of a special relationship that Niels Bohr called *complementarity*.[1] One can think of the complementary relationship between two observables as the two sides of a coin: e.g., momentum is heads and position is tails. The Heisenberg Uncertainty Principle (HUP) dictates that you cannot see both of the sides at once, but they are both aspects of the coin. It turns out that a similar relationship applies to measurements of time and energy, although time itself is not really an observable in quantum theory. The HUP can be applied to time and energy just as it can be applied to position and momentum, although (since time is not an observable) it is not obvious exactly what 'measuring time' means in physical terms, and there is some debate about it among physicists. Below, I'll discuss what it means in the transactional picture.

Recall that the HUP tells us that if we are measuring an object's position very accurately, its momentum becomes very uncertain, and vice versa. One way to understand the application of HUP to time and energy is by recalling that time is really just a measure of change: the 'passage of time' corresponds only to an observable change or set of changes, as in the difference in the states of the clock of Figure 8.2. The HUP, when applied to time and energy, gives us a rather striking result: it tells us that if we have zero uncertainty in an object's energy, there is a complete lack of any change in the object that would allow us to apply any definite time interval to that object other than the one defined by the creation and destruction of

[1] Bohr first proposed this concept in 1928.

the object itself. That is, not only its energy but also no other observable properties of the object can change, because any such change would imply the passage of a well-defined, finite time interval, just as the changes in a clock reflect well-defined, finite time intervals.

If a well-defined time interval were applied to an object with well-defined energy, this would contradict the HUP, which says that there must be infinite uncertainty in any time interval applying to any object with a perfectly well-defined energy. This boils down to saying that any time interval applying to the object cannot be distinguished from the lifetime of the object. The object would be like a clock that winks in and out of existence, and during its entire lifetime its hands do not move. In other words, such an object would make a very ineffective clock, because it could not define any time intervals!

Consider now a quantum other than a photon, with a definite energy including a non-zero rest mass (such as an electron emitted during radioactive decay, as in our discussion of Schrödinger's Cat in Chapter 5). Suppose, for simplicity, that there is only one absorber available to it. When it is conveyed from an emitter to that absorber in an actualized transaction, its lifetime is measured by its emission and absorption. The HUP tells us that because it has a definite energy, there can be no smaller (i.e., more well-defined) time interval that applies to it. We cannot divide its lifetime into any smaller intervals of time. As far as the electron is concerned, it came into existence and then it went out of existence, and that is all. This is in contrast to the ordinary clock above, whose changes record many distinct time intervals, all of which apply to the clock. That is, the clock's existence can be subdivided into many time intervals, while the quantum with exact energy cannot.

Another way to understand this point is to think in terms of a movie. The electron offer wave, having a definite possible energy, is emitted by atom A and confirmed by atom B. The resulting actualized transaction establishes only two frames in the film clip, from the electron's point of view. There is a single, indivisible time interval, and its beginning and end are defined by the two frames (see Figure 8.3).

Thus, the most accurate way to understand the time-energy version of the HUP is not really in terms of time intervals, but rather in terms of change: if an entity has a completely precise energy value, there is no

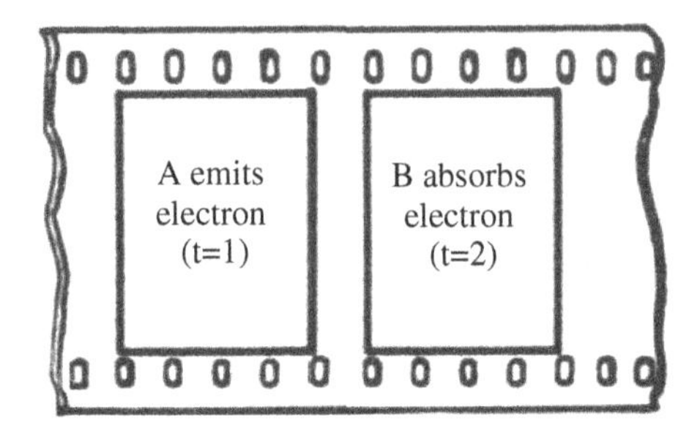

Figure 8.3. A film clip of an emitted and absorbed electron has only two frames.

observable change in the entity characterized by that energy. However, there are changes in the emitter that emitted it and in the absorber that absorbed it: one lost energy, and the other gained it. The clear distinctions between the 'before and after' states of the emitter and absorber define two events; this is what establishes the two identifiable frames of Figure 8.3. This, in turn, allows us to define a time interval corresponding to the two events. This leads to the crucial point that a time interval is created by the transfer of some exact amount of energy from one entity to another in an actualized transaction. The entity that delivers that energy — the actualized quantum — is an indivisible, singular process that cannot be broken down in any finer detail than by the emission event and the absorption event in which it participates.

In terms of the film clip representing the transaction in Figure 8.3, this means that there cannot be an arbitrarily large number of individual frames, and they cannot be arbitrarily close together. If your film clip includes the emission and absorption of a particle in an actualized transaction, you cannot ask to see the frame for the process at some time in between the particle's emission and absorption. As far as the particle is concerned, there is no such time.

Behind the scenes: Film clip production

Before going further with the idea of a film clip, we must acknowledge that a 'film clip' implies a spacetime process. Yet, much of the transactional process does not occur within spacetime. So, in order to see what's going on, we must imagine the behind-the-scenes production of the film in question. Consider again the two frames of the clip depicted in Figure 8.3.

What is depicted there is the actualized transaction, in which a quantum of real energy is delivered from the emitter at $t = 1$ to the absorber at $t = 2$, creating the time interval that has its beginning and end defined by the two frames. But a lot of production work goes on behind the scenes, in Quantumland, in order to create this film clip. The emitter A sends out an offer wave, and the absorber B confirms it. Perhaps there are competing confirmations from other absorbers, but absorber B 'wins' the competition, actualizing the transaction. It is only at that point that the film clip is produced. That is, the two distinct frames are only brought into being only as a final result of the transactional process, which results from the behind-the-scenes negotiation between the emitter and absorber(s). All of this negotiation was conducted outside spacetime. Remember, spacetime is just the collection of all actualized events, and those events originated from Quantumland.

This naturally leads us to wonder: where and when are atoms A and B during the production process? Again, they are not 'in spacetime' at all; they are in 'Quantumland.' As we discussed in the previous chapter, atoms A and B are not localized in spacetime until a transaction occurs and a spacetime interval between them is created. They are the actors, if you will, helping to create the movie. Only if they engage in a transaction (i.e., the cameras are rolling and a scene is being recorded) is a spacetime interval established between them, and that interval is recorded in the film clip. We will examine an additional aspect of this production process in the next section.

Generators of space and time

We've noted above that, in the transactional picture, a time interval is created by the delivery of a quantum of energy from an emitter to an absorber. Interestingly, this corresponds to a more esoteric but standard physical account of the relationship between energy and time: Energy is the creator of time intervals.[2] The relationship of energy with time has an exact mathematical representation, but we don't need that for our purposes. In ordinary language, as we've already noted in the previous section, it just

[2]Technically, it's 'energy is the generator of time translations,' where a 'translation' is a displacement from one coordinate index to another.

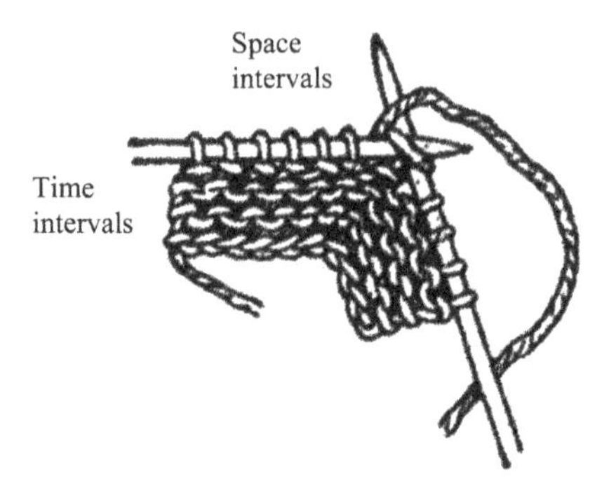

Figure 8.4. The fabric of spacetime with spatial and temporal intervals indicated.

means that a quantity of energy corresponds to the passage of an interval of time; without that quantity of energy, there is no passage of time. One way to visualize this concept is to recall from the previous chapter our picture of the 'fabric of spacetime' as it is extruded in a metaphorical knitting process (see Figure 8.4). The different stitches on a given row, horizontally across the row, can be compared to different positions in space.[3] The different rows, stacked vertically, can be compared to different times. In these terms, it is energy transfer that creates each new 'row' of the knitted fabric: the transferred energy is the 'yarn' that separates one row from another.

At this point, you may have guessed that a similar relationship holds for position and momentum. Indeed, it's also true that momentum is the creator of spatial intervals. As with energy and time, a transferred quantity of momentum (basically motion) corresponds to an interval of space; and without that momentum, there is no spatial interval. In transactional terms, a spatial interval is created by the delivery of a quantum of momentum from an emitter to an absorber. In terms of the knitting metaphor, the yarn that goes between stitches on the same row is like a momentum transfer. Without that yarn, there is no separation between stitches; i.e., no spatial interval.

[3]Although our knitting analogy has several stitches on the needle at the same time index, this is not to imply that they are all simultaneous for everyone. As we have seen previously, time and space are relative to the state of motion of the person doing the measuring.

Let us further explore this idea of the creation of spatial and temporal intervals by momentum and energy transfers. We first need to recall the de Broglie wave, which is the basis of both offer and confirmation waves. The de Broglie wave has a frequency and a wavelength; the energy is defined by the frequency of the wave, and the momentum is defined by the wavelength of the wave.[4] But recall that these are only *possible* energy and momentum. Now, suppose an excited atomic electron emits a photon offer with a given possible energy and momentum (frequency and wavelength). Another electron capable of absorbing that energy and momentum responds with a confirmation, and a transaction is actualized. The momentum transferred in the actualized transaction (given by the wavelength of the offer and confirmation) is what creates the spatial separation between that emission and that absorption. It is important to recall from the previous chapter that the spacetime events corresponding to the emission and absorption are brought into being by the actualized transaction; they did not exist before.

However, relativity tells us that neither spatial nor temporal intervals, individually, are absolute quantities. They are defined only relative to the states of motion of observers who can see the emission and absorption events marking those intervals. And the quantum's amount of motion itself is not absolute; that is, whether or not a quantum has momentum depends on the frame in which it is described. For example, if we think of ourselves as an actualized electron being delivered from an emitter to an absorber, from our point of view the absorber is coming towards us, while we are at rest. To picture this, think of the absorber as somewhat like a bus arriving to pick up a rider (the actualized electron) waiting at a bus stop.

Nevertheless, there *is* an absolute quantity, one which is the same for all observers regardless of their relative motions. It is the spacetime interval, mentioned in the previous chapter. The spacetime interval is basically the difference of the (squares) of the temporal and spatial intervals as measured in any particular frame. (You take the time interval between events A and B, multiply that by the speed of light, square it, and then

[4] For readers who would like more technical detail, the energy is the frequency multiplied by Planck's constant h, and the momentum is h divided by the wavelength.

subtract from that the square of the distance between A and B.) It turns out that despite the differing perspectives of observers in different states of motion, the spacetime interval is the same in all of them. It is the one measurable spacetime quantity that they can all agree on, even though the individual space and time intervals depend on the reference frame. And the spacetime interval is what is created in an actualized transaction.

Rest frames

Above, we noted that a spatial interval is created by momentum transfer. But what about the frame in which the offer wave's momentum is zero? This frame is called the rest frame.[5] This is your frame of reference when you are cruising in a plane at a constant speed; you do not feel motion, and the clouds seems to be going past you. To address the situation in the rest frame of the quantum, we also have to note that in relativity, a quantum's mass depends on its velocity. The higher the velocity, the greater the mass of the quantum. The mass of a quantum in its rest frame is called its rest mass; that is, the smallest mass a quantum can have. But there can be a rest frame only for a quantum with non-zero rest mass. A photon, whose rest mass is zero, does not have a rest frame: there is no frame in which a photon can be considered at rest.[6] We'll return to the spacetime implications of the massless photon a bit later.

Consider now a transaction involving a quantum with some rest mass, but in its rest frame, in which it has zero momentum (i.e., zero velocity). In that rest frame, no momentum is transferred, so there is no spatial interval defined in that frame. That is, relative to its rest frame, the quantum actualized in such a transaction does not travel any spatial distance. As seen in that frame, it is stationary. Returning to our bus analogy, it is like the rider at the bus stop being dropped off by one bus (the emitter) and waiting for a connecting bus (the absorber) to pick it up. But in the quantum's rest frame, where its momentum is zero, it still has some energy: this is called rest energy, and it is due solely to the quantum's rest

---

[5]This means that its wavelength is infinite. However, it can still have energy (rest energy), so its frequency is not zero.

[6]This means that from the photon's point of view, there is no spacetime.

mass. The rest energy is just the rest mass multiplied by the square of the speed of light (remember that $E = mc^2$). This rest energy is what is transferred from the emitter to the absorber as seen in the rest frame of the quantum, and it generates an interval of time (but no space) as seen in that frame.

The Inequivalence of Space and Time

As noted in the previous chapter, even though measurements of space and time are relative, the dimensions of space and time are not interchangeable. It turns out that (from the point of view of its rest frame) a quantum can create time without space, but it cannot create space without time. The reason is that a quantum can have energy without momentum, but it cannot have momentum without energy. As noted above, rest energy is the energy that a quantum can have without motion (or momentum); and that rest energy can be transferred. Since a time interval is created from energy transfer, a time interval is created whether or not the transferred quantum has motion. On the other hand, if a quantum is moving, it has momentum, and therefore it creates space; but it also has energy of motion, so it also creates time. So a necessary byproduct of the creation of space through momentum is the creation of time. This is because there is always an amount of energy associated with motion; that is, you must have energy in order to be in motion.

The restless photon

We can better understand the relationship of matter to the generation of time by considering a quantum that has no mass: the photon. As I noted above, the photon has zero rest mass, and therefore there is no rest frame for the photon; i.e., there is no frame in which the photon could be at rest and therefore would have zero momentum. So the photon is always in motion, and therefore always transfers momentum along with any energy that it transfers. Because it has no mass, it can never transfer rest energy, but only kinetic energy. So (from the standpoint of another object with a rest frame) a photon always creates both a spatial interval and a time interval. But in the case of the photon, it turns out that these are always exactly equal. Recall that the spacetime interval is the temporal interval squared minus the spatial interval squared. Since the spatial and temporal intervals are always equal for the photon, the spacetime interval for the photon is always zero (as measured in all reference frames).

In this situation, relativity dictates very interesting consequences for both time dilation and length contraction (we mentioned these in the previous chapter). If you think of the photon as leaving a clock face, to the photon the clock appears frozen, so there is no passage of time from the photon's point of view.[7] Meanwhile, the spatial interval is infinitely contracted, which means that the distance between the photon's emission and absorption has been shrunk to zero, from its perspective.

Thus, it turns out that, according to relativity, the photon itself experiences neither the passage of time nor space: as far as it is concerned, it has not travelled any distance, or has it experienced any passage of time. This is all because the photon has zero rest mass. But the minute you add a tiny bit of rest mass — no matter how tiny — space and time are born (from the point of view of the object with mass). It turns out that time is more fundamental for a quantum with rest mass, since that quantum has a rest frame and therefore can generate an interval of time without necessarily generating an accompanying interval of space. Space and time are therefore distinct features of the physical world, and are not interchangeable. In fact, you can think of rest mass as the substance that generates time, since rest mass is what defines a rest frame in which only an interval of time is generated. Rest mass is figuratively the 'sand of time' in the hourglass (Figure 8.5).

Energy vs. momentum

There is yet another reason why space and time are not interchangeable. For any transaction that actualizes a quantum, there is always a delivery of a quantum of positive energy from the emitter to the absorber, never a delivery of negative energy from the emitter to the absorber. In other

[7]This was actually a 'thought experiment' explored by the young Albert Einstein prior to his development of the theory of relativity.

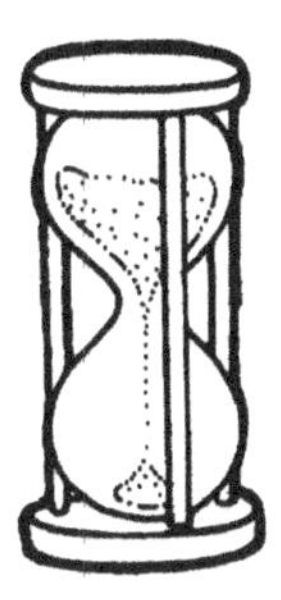

Figure 8.5. Matter is the 'sand in the hourglass': you need rest mass in order to experience the passage of time.

words, energy is always conveyed in such a way that the emitter loses positive energy and the absorber gains it. In contrast, an emitter can emit negative momentum (rather than energy) and an absorber can gain negative momentum. This is because 'negative momentum' is just travel in the opposite direction from the 'positive' direction, and any direction of spatial travel is possible. Indeed, the choice of which is considered the 'negative' direction and which is considered the 'positive' direction are entirely arbitrary (see Figure 8.6). For technical reasons, it turns out that the reversibility of momentum in this way is the reason that its complementary property, position, qualifies as an observable, while the nonreversibility of energy is the reason why its complementary property, time, does not qualify as an observable.[8]

But the reader might ask: why don't we have the same situation with time and energy? That is, why can't a ground-state electron 'emit' a quantum of 'negative energy' and thereby transition to a higher energy state? In that case, the 'negative energy' would then be 'absorbed' by an excited state electron, which would then transition to its ground state. The answer

[8] But position qualifies as an observable only in the nonrelativistic theory. In the relativistic theory, no spacetime quantity, such as position or time, is an observable; only dynamic quantities such as energy and momentum are observables. Since the relativistic theory is the more general and accurate theory, this indicates that these dynamic quantities are more fundamental than spacetime quantities. This point corroborates the picture presented here: namely, that spacetime is emergent from the quantum level, which creates energy and momentum transfer.

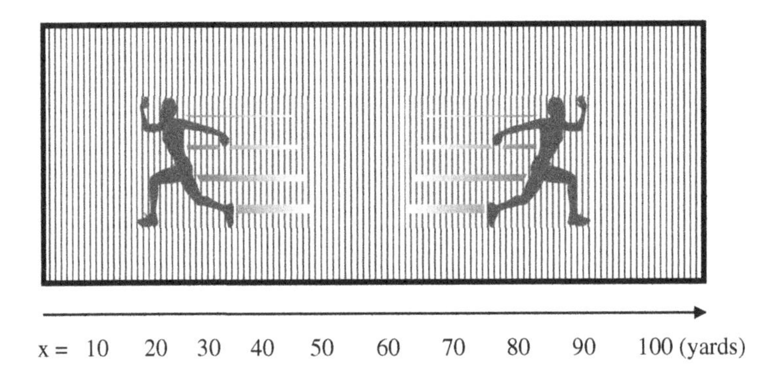

Figure 8.6. Two runners run toward opposite ends of a field. According to the convention shown, the runner going to the right has positive momentum, because he is running toward higher numbers in this system, and the one going to the left has negative momentum. But if we decided to count the number of yards from the right to left instead, these descriptions would be reversed. There is no fundamental physical significance concerning which direction is considered positive and which is considered negative.

is that you cannot do this, because 'absorbing negative energy' is exactly the same thing as emitting positive energy. That is, if an atom were to 'emit negative energy,' it would be gaining positive energy, so it would be an absorber, not an emitter. Remember that offer waves always have positive (possible) energy, while confirmations always have negative (possible) energy. For an atom to 'emit negative energy,' the confirmation would have to be the starting point for the transactional process, and that's not how it works, because confirmations are always responses to positive-energy offer waves.

Another way to see this is in terms of a financial transaction, where we compare energy to money. Income is positive energy, and debt is negative energy. You can have income added to your account, or you can have a debt removed. The first is analogous to your gaining positive energy, and the second is analogous to your emitting negative energy. They have exactly the same effect on your bottom line: it has increased. This means, in effect, that you have absorbed money.

In contrast, as noted above, an emitter can emit quanta having either positive or negative momentum, and neither of those processes is equivalent to absorbing anything. Recall also that relativity tells us that you can generate a temporal interval without any spatial interval (in your rest frame), but you cannot do the opposite. You cannot turn time into just another spatial dimension. Time is physically distinct from space, just as energy is physically distinct from momentum.

The Now and Temporal Direction

In this chapter, we've talked about the unidirectional 'flow' of time. But recall from the last chapter that the Now is where all the action really takes place, and the actualized events constituting what we call 'spacetime' are extruded from that. If there is any 'flow,' it is really the flow of actualized events receding away from us, into the past. Our sense of 'moving into the future' is really the perception of the changes in the stitches on the needle, metaphorically speaking. The Now, which is the entire field of our experience, does not move anywhere (or rather, anywhen); rather, it changes. The changes are brought about by transfers of energy from emitting entities to absorbing entities, and time is just the attribute by which we measure those changes. In this sense, energy and time are complementary aspects of the same thing; that is, they are 'two sides of the same coin': change (Figure 8.7).[9]

The emission of a positive energy offer wave is always the starting point of the transactional process. An offer of positive energy must be created (emitted) before it can be responded to by a negative-energy confirmation. When a particular incipient transaction is actualized, a quantum of real energy must be made available by the emitter before it is destroyed at the receiving absorber.[10] In both cases, energy is created before it is destroyed, and the created energy is always positive.[11] This asymmetry in creation vs. destruction is what gives directionality to the process of change, and therefore to the attribute of time by which that change is measured.

---

[9] Is it a coincidence that coins are called 'change'? In any case, this recalls the point made earlier about time and energy being complementary.

[10] Technically, absorption is expressed as destruction in relativistic quantum theory.

[11] The reader who knows about anti-particles may wonder about the standard account of an anti-particle as a 'negative energy particle.' But, in fact, any actualized antiparticle always conveys positive energy from an emitter to an absorber, and an antiparticle offer wave is also characterized by positive (possible) energy. The characterization of an antiparticle as a negative energy particle is only truly appropriate at the virtual quantum level, where no emission or absorption is really defined. Any actualized quantum is always a positive-energy quantum, and therefore always establishes a forward temporal direction.

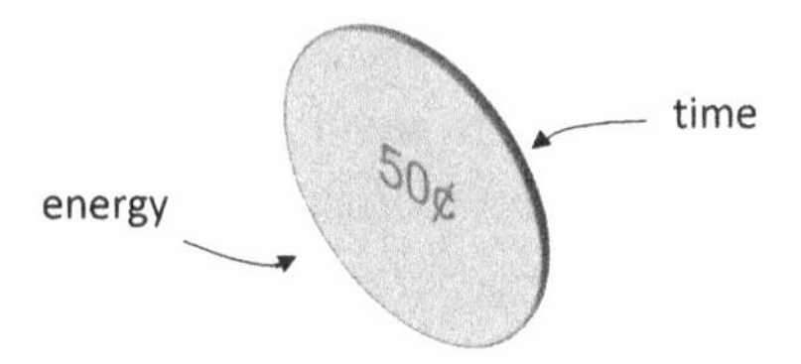

Figure 8.7. Energy and time are 'two sides of the coin' of change.

The garden of forking paths[12]

An additional aspect of the directionality of time can be found in terms of the lightning strike with which we began this chapter (Figure 8.1). The lightning strike begins at one point, but it can, and often does, branch to land at more than one point. In the same way, a photon is emitted by one entity (such as an excited atomic electron) but is generally confirmed by many potentially-absorbing entities (such as ground-state electrons). But, in contrast to the lightning strike, which can distribute its electrical energy to many points on the ground, a quantum emitter can give a quantum of energy to only one of the responding absorbers, since quanta are indivisible (this is why they are commonly pictured as 'particles'). As discussed in Chapter 3, this indivisible quantum of energy is conveyed by a single transaction that is actualized from a set of incipient transactions. This process of actualization corresponds to the conventional quantum mechanical notion of collapse, and it is genuinely unpredictable. The intrinsically-unpredictable and many-to-one quality of collapse is what makes the future 'open,' or indeterminate. The availability of more than one absorber for any given emitted offer wave, and the confirming responses of those absorbers, is what creates a 'garden of forking paths,' only one of which will actually be realized. Thus, we have identified two important aspects of temporal asymmetry: (1) creation (emission) precedes destruction (absorption); and (2) the emitter is singular, while its possible absorbers are multiple. Point (2) is what makes the future uncertain. Let's now consider (2) in a little more detail, to see where this uncertainty comes in.

---

[12]This is the title of an intriguing short story by Jorge Luis Borges (1941, English translation by Boucher, 1948).

In Chapter 4, we noted that virtual photons are being tossed around all the time by charged quanta such as electrons. Remember that there is no fact of the matter about the emission and absorption of these virtual photons, because they are not really being emitted and absorbed. However, recall that a virtual photon may be spontaneously elevated to a photon offer wave under the right circumstances and by 'beating the odds' related to the fine structure constant (1/137). The elevated photon offer wave acquires a clear emission point, and it turns out that, in general, it will have a multiplicity of available absorbers, all of which will respond with confirmations. The existence of such a multiplicity of responding absorbers is based on an interesting feature of the state of motion of the photon offer wave: the offer wave goes in all directions at the same time, like the circular waves that ripple outward from a stone thrown into the water.

To explore this issue further, we need to recall from Chapter 2 that momentum is the physical quantity that tells us how much motion is present, and where that motion is directed. Momentum can be represented by an arrow; the length of the arrow is the magnitude of the momentum (meaning the amount of motion, or speed), while the direction of the arrow indicates the direction of the momentum. (See Figure 8.8 for some illustrations of different momentum values.)

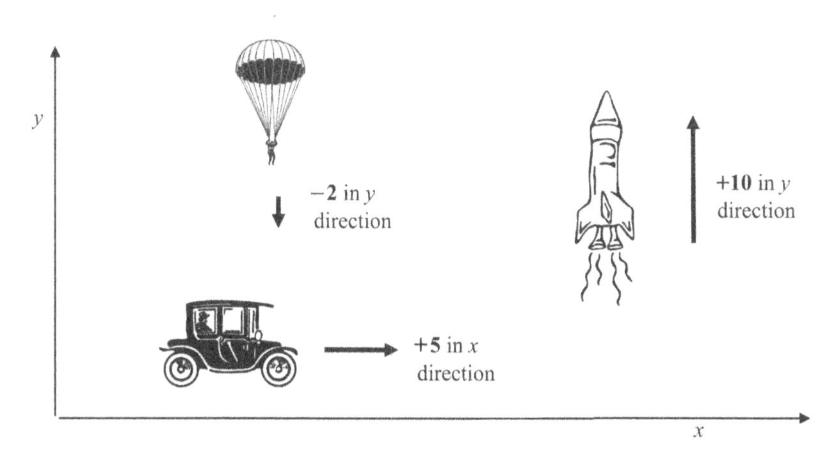

Figure 8.8. Illustration of momentum values. The size of the arrow gives the amount of momentum, while the arrow's direction tells us the motion's direction, including whether the momentum is positive or negative with respect to a given direction.

Returning now to a photon offer wave: it turns out that only the magnitude of its momentum — the amount of motion but not its direction — is well defined. What this means is that it is a spherical offer wave, emitted in all directions at once, as illustrated in Figure 8.9. Therefore, while a virtual photon is tossed symmetrically between two quanta in a one-to-one relationship, a photon offer wave is the starting point for an asymmetrical physical process; a one-to-many relationship. This is because it becomes accessible to many potential absorbers, owing to its lack of commitment to any particular momentum direction. This is indicated in Figure 8.10, which shows many absorbers as black dots, each receiving only one of the arrows representing a specific directional component of the photon offer wave. Each of those absorbers responds with a confirmation corresponding only to the momentum component that reached it. In this way,

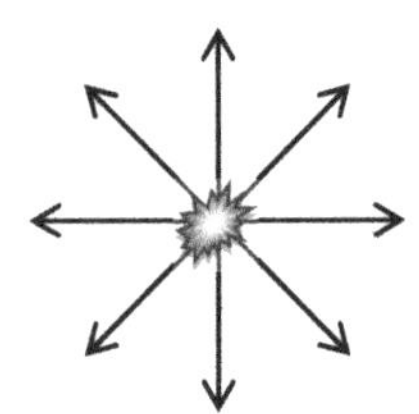

Figure 8.9. A photon offer wave is omni-directional.

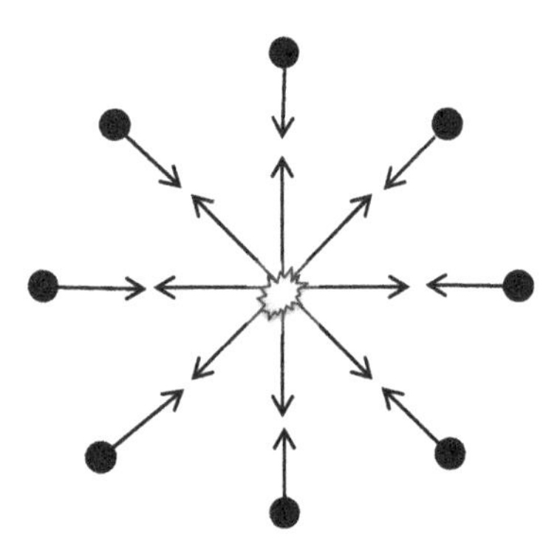

Figure 8.10. Many absorbers respond to a single omni-directional photon offer wave, each picking out a particular momentum direction and responding with a confirmation for that direction only.

the photon offer wave's lack of a preference for any momentum direction, together with the availability of many absorbers capable of picking out a particular direction, creates a 'garden of forking paths.'

As a result, what we get is a set of incipient transactions, each corresponding to a different momentum direction. Only one of these can be actualized, and when that happens, a real quantum of electromagnetic energy with a particular momentum direction is delivered from the emitter to the 'winning' absorber. In short, that momentum direction has now been actualized where none existed before. A single direction has been chosen, where many directions were possible.

Symmetry breaking creates temporal direction

Recall for a moment our discussion of symmetry breaking in Chapter 3; this is the process by which one incipient transaction is actualized out of many possible ones. The basic phenomenon of symmetry breaking is also how the highly symmetrical character of the virtual particle exchange is transformed into the above highly asymmetrical process of emission and absorption of an offer wave.

Recall also from Chapter 4 that a virtual quantum can be 'off mass shell.' That is, it is not constrained to the Einstein relation $E = mc^2$. In this state of existence, it cannot satisfy the conservation laws that apply to any observable, spacetime process, and therefore it cannot participate in an emission and absorption that would create a pair of spacetime events. However, coupling between charged particles can give rise not only to virtual photons that are not on the mass shell, but also to photons that are on (or very close to) the mass shell. Such a photon may arise in a strongly asymmetrical relationship among charged particles, for example between an excited atomic electron and one or more ground-state electrons. If the energy of the candidate photon is equal to the difference between the electron's excited state and an available lower state, the photon acquires a well-defined emission point: namely, the excited electron. Of course, in order to actually be emitted as an offer wave, that photon must survive the 1/137 odds (as well as the other hurdles discussed in Chapter 5). If it does not, it remains a virtual photon, which can contribute to force but which cannot participate in a transaction and thereby transfer real energy.

Now, suppose that all the relevant odds are surmounted, and the photon is emitted. Recall that, while it may be on the mass shell, such a photon still does not have any preferred momentum direction. If there is more than one (ground-state) electron available to it, the multiplicity of that set of absorbers sets up a competition in the form of a set of incipient transactions. Each of those is what defines a momentum direction, as in Figure 8.10. This is how spatial directionality is brought into being in the first place: it stems from the one-to-many relationship that typically occurs when a real photon is brought into being.

If there were only one excited-state electron and one ground-state electron in the whole universe, the photon would be actualized as a quantum of electromagnetic energy delivered from the excited electron to the ground-state electron, and they would simply swap roles. There would be only one spatial direction generated by such an exchange. That one direction corresponds to only one spatial dimension. Therefore, the spacetime realm based on that one transaction would be a one-dimensional spatial world, much like 'Lineland' from the *Flatland* story.[13] In order to have a three-dimensional world, we need at least three absorbing atoms (so that three perpendicular directions are established).

Time's Arrow Needs True Randomness

In the previous section, we considered the way in which a 'garden of forking paths' is created through the transactional process. This process is truly indeterministic: there is no way, even in principle, to predict where the emitted quantum will end up, even if we know its emission point. When the quantum of real energy is absorbed at the receiving absorber in an actualized transaction, a truly random process has occurred. This sort of randomness is apparent in many observable processes, so it is generally taken for granted and even assumed as a crucial part of the explanations for many everyday phenomena. Those explanations take the form of

[13] 'Lineland' is a hypothetical realm discussed by Abbott (1884) in *Flatland*, composed only of a single spatial dimension. Its inhabitants are simply points on the line.

statistical arguments, which apply probabilistic calculations to processes assumed to be fundamentally random.

However, you cannot get this crucially-necessary randomness from any of the deterministic laws of physics, not even from those applying at the quantum level. Only with an indeterministic collapse, as in the transactional interpretation, do we see true randomness. (Recall from Chapter 2 that traditional approaches to quantum theory recognize only a deterministic law that applies to the way the quantum possibilities interact with a measuring device or with each other. The deterministic behavior of the quantum state is represented by the 'Schrödinger Equation'.) This puzzle even tripped up Ludwig Boltzmann, mentioned earlier in connection with his brilliant atomic theory (Boltzmann, 1896). Specifically, Boltzmann tried to derive the observed irreversible phenomena of dissipative processes (such as cream mixing into coffee) from the deterministic laws that governed the behavior of the atoms in his theory. He was partially successful, but only because he 'smuggled in' the idea of the random motion of those atoms via his statistical analysis. He did this by assuming that atoms, and the molecules they comprised, were engaging in 'chaotic' motion.

Boltzmann was right: there really is what he termed 'molecular chaos,' even though it is not derivable from the deterministic laws that he assumed applied to his atoms. We see this 'chaos' in the form of the random thermal motions that give rise to what is called Brownian motion in the modern era. Indeed, these sorts of thermal motions were observed millennia ago and, even in those ancient times, attributed to atoms! Here's a lovely account by the Roman poet Lucretius (c. 99–c. 55 B.C.), from his poem 'On the Nature of Things':

> Observe what happens when sunbeams are admitted into a building and shed light on its shadowy places. You will see a multitude of tiny particles mingling in a multitude of ways [...] their dancing is an actual indication of underlying movements of matter that are hidden from our sight [...] It originates with the atoms which move of themselves [...] the movement mounts up from the atoms and gradually emerges to the level of our senses, so that those bodies are in motion that we see in sunbeams, moved by blows that remain invisible. (Lucretius, 2008)

Those 'invisible blows' are the collisions of the dust particles with truly random motions of the air molecules. These motions are called 'thermal' because they are attributed to heat. What is heat? It is just energy that has been randomized so that we cannot say where it came from or where it is going; it is not subject to deterministic, predictable laws.[14] In the present interpretation, this energy is transported randomly from molecule to molecule by thermal photons in actualized transactions. Thus, the transactional picture also explains the truly random nature of heat, which is otherwise still a mystery. (Thermal photons have wavelengths in the 'infrared' range of the electromagnetic spectrum, pictured in Figure 2.2.)

In Chapter 6, we saw that macroscopic objects are sustained as coherent spacetime objects by these sorts of ongoing transactions among their constituents. Now we see that it is the random collapse of the transactional process that also provides the crucial indeterminism beneath the irreversible macroscopic phenomena with which we are so familiar, such as cream mixing into coffee but not out of it. These kinds of irreversible processes indicate a clear direction for time's arrow; they allow you to say why the future is different from the past.

Free Will

The indeterministic 'garden of forking paths' that is created each time an offer wave is emitted has important implications for the concept of free will. Free will is generally understood as the idea that an agent (such as a human being) has free, uncompelled choices as to how to act, and that he or she can implement those choices by interacting with the physical world in a way that brings about the intent of the chosen actions. The two main features of free will are: (1) the choosing agent is not fated to encounter already-defined events in the future (this is the idea that the agent is faced with a genuine 'garden of forking paths'); and (2) the agent has the ability

[14] Heat has to be truly indeterministic to give rise to the macroscopic laws of thermodynamics. The origin of this indeterminacy remains an outstanding problem in physics — except in the transactional picture, where it comes from the indeterministic collapses that actualize one transaction out of many incipient ones.

to exert his or her volitional force on physical objects in such a way as to implement those choices.

There is an additional element to free will: the concept of volition itself, the origin of the external agent's choices according to intention, desire, or other considerations. This subject is outside the realm of physics, and the interpretation presented in this book has nothing specific to say about it.[15] The main point being made in this section is that, given volition, an agent can freely act on that volition in an effective way.

Requirement (1) is fairly straightforward: to be truly free to choose, we must not be predestined or fated to encounter future events.[16] Requirement (2) is that the agent has a conscious volitional capacity that is external to the processes underway in the physical world, and which allows the agent to intervene in those processes. Metaphorically speaking, one can think of those physical processes as a river following its course. A volitional agent in the sense of requirement (2) can be thought of as a person who comes along and diverts part of a river with an aqueduct for some purpose (such as irrigation). We need to be able to intervene in the river's flow (i.e., the flow must not be fated to continue unaltered). But we also need to be able predict what will happen as a result of our intervention; we intend for the redirected flow to go in a specific direction, not in many different directions, or in some random direction that is a surprise to us. Free will is meaningless if an agent has the freedom to choose how to act but has no idea what the results of his or her actions are going to be.

As noted above, the indeterminacy concerning which transaction will be actualized naturally leads to a physical basis for the fulfillment of requirement (1) of free will. Specifically, in this picture, the future is genuinely undetermined because it is composed of unactualized

---

[15]This remains a topic for further exploration, however. Could virtual quantum activity have some relationship to the origin of volition? Could nature's 'choice' of which transaction to actualize be based on volition? In general, does the mental realm have its origin in quantum possibilities? There are all legitimate philosophical questions worth pursuing.

[16]Even though this seems straightforward, there are many philosophers who argue that one can be completely fated and predestined and still have free will. That position is called 'compatibilism.' I leave it to the reader to decide for him- or herself whether one could have genuine free will while at the same time being unavoidably fated to make all the 'choices' one will ever make. More details on this topic can be found in Appendix B.

possibilities. That's what quanta are. Recall from the knitting analogy of the previous chapter that quanta are the underlying reality, the metaphorical 'yarn' that gives rise to spacetime events through actualized transactions. Remember also that the 'Now' is the set of stitches on the needle, and the past is the extruded set of previously-transacted events. This means that there are no 'future events,' already sitting there in spacetime and which we are inevitably fated to encounter. Instead, there are many different possible outcomes for any given situation that may develop in the 'knitting' process, but only one of these possibilities becomes actualized. So in this picture, it appears that we are not subject to fate, as we would be in a block world. This means that the transactional picture satisfies the first condition listed above: we are not fated to encounter specific events. We really do have a 'garden of forking paths,' and therefore we have genuine choices.

Therefore, requirement (1) is clearly satisfied in the interpretation presented here. Nevertheless, one problem that has faced traditional attempts to obtain a consistent account of free will is that requirements (1) and (2) seem to be at odds with each other. Here's why: requirement (1) says that to have genuine choices, we need the future not to be fated. This means that the governing laws have to be indeterministic, which means that a particular initial action does not have a single predetermined result. Instead, it could lead to many different results. This implies that the results of an action are not predictable. But this seems to be in conflict with requirement (2), since that requires that an agent be able to intervene in processes in a reasonably predictable way, in order to achieve a chosen goal.

We are therefore presented with a dilemma: we need indeterminism to have free choices, but we need determinism in order to meaningfully act on those choices. How can we escape this dilemma? The short answer is that in order to have genuine free will, we need both quantum processes and classical processes. Free will is the dynamic field of action in which the quantum and classical worlds confront each other. In what follows, we'll see how this might work.

First, note that quantum processes have the indeterministic character that is needed for requirement (1), while classical process have the law-like, predictable character that is needed for requirement (2). At the quantum level, individual events are not predictable; they are genuinely indeterministic. But, as possibilities of varying strength, they do have tendencies, and those tendencies result in probabilistic laws that become manifest when large numbers of quantum systems are involved. However, the basic hallmark of quantum indeterminacy is that for a given individual quantum system, we cannot in general predict what will happen to that particular system at a future time. The system seems to 'fork' into a set of possibilities, a garden of forking paths. In contrast, the classical, macroscopic world is deterministic: for each input, there is only one output. Given an individual macroscopic system, such as a batted baseball, we can predict its future course with great reliability. The only uncertainty about that prediction is due to limitations in the accuracy with which we knew its initial state.

In Chapter 6, we discussed how that macroscopic, classical world naturally emerges from the quantum realm. This emergence of classical predictability is due to the collapsing effect of absorptions, which are virtually guaranteed by large enough collections of absorbing quanta. This process is going on all the time; it is what creates the classical world of experience that we call 'spacetime.' For example, imagine that you are a spectator at a baseball game on a sunny day. The pitcher's baseball is created as a localized object by continual transactional collapse due to energetic interactions among the quanta that comprise it (intra-baseball transactions, if you will) and between the baseball and external quanta (inter-object transactions). An example of such an external object is the Sun, which emits photon offer waves that are absorbed by eligible atoms in the baseball, resulting in actualized transactions that localize both the quantum that emitted the photon (in the Sun) and the quantum that absorbed it (in the baseball). Once you have localized objects such as this, the laws of classical physics apply, and you can predict the behavior of the objects with great reliability.

Thus, the fabric of spacetime — that is, the set of actualized events — is continually being created from the transactional process at the inherently unpredictable quantum level. The predictability of classical physical laws emerges only at the macroscopic level of actualized spacetime events. At the quantum level, the world is genuinely indeterministic. And this is where our free choices can enter. Thus, we can fulfill both requirements (1) and (2), but at different levels of reality, so that they do not have to conflict with each other.

Another component of the solution to the free will dilemma is to make a distinction between mind and matter. If our minds operate (at least in part) in the quantum realm, which is not part of spacetime, then our thoughts are not subject to classical predictability. This would seem to be a reasonable conjecture, since thoughts and other forms of mental activity are not spacetime phenomena.[17] Yet they are experienced as real by conscious agents, at least on an internal level, and they can certainly lead to concrete interactions with the physical world that bring about specific empirical events. So, might thought itself dwell somewhere in Quantumland?

In fact, drawing a distinction between mind and matter has been rather traditional in philosophical thinking. Descartes was one famous philosopher who embraced such a view, known as *dualism* (Descartes, 1641). However, the suggestion here that thought might be governed by quantum physics is a modern variant of this traditional separation between mind and matter. In exploring this idea, we venture beyond the usual domain of physics and into the domain of philosophy. This is because (at least at our current state of knowledge) thought is not something that is amenable to physical description, at least not in a way that could be empirically corroborated. However, there is certainly a logical thread that links this idea to physical theory: as noted above, quantum entities are not contained in spacetime, and they function as possibilities. Similarly, thought is not contained in spacetime, and it deals with possibilities, especially if we consider that thoughts precede actions, and we can consider a variety of different actions using thought.

The suggestion here is that our basic freedom of choice arises at the mental level, which is not constrained by the deterministic laws of classical physics. (Note that we are not talking about the physical brain, but

---

[17] It is known that certain kinds of mental activity can be traced to electrical impulses in the brain. However, these empirical phenomena are distinct from mental entities such as thoughts and ideas. That is, a thought or idea is not reducible to electrical impulses.

rather the intangible level of thought itself.) The mental level may well be described, at least in part, by quantum processes, which would allow it to interact effectively with the types of quantum entities discussed in this book and which underlie our observed physical reality. Meanwhile, that classical level of appearance only emerges due to the actualized transactions which create the 'fabric' of spacetime. So our opening for free will is contained in the transactional process; that is, in the indeterministic emission of offers and confirmations, and the indeterministic realization of one of the resulting incipient transactions. In this way, the transactional picture of quantum reality could break new ground to a fruitful new avenue for researchers of the future to gain a better understanding of the mind, including its volitional capacities, which has been a topic of questioning and research for millennia.

Finally, I should note that the philosophical problem of free will is huge and vast, and there are many different points of view on the subject. In Appendix B, I summarize some of those views, along with a brief sketch of how the interpretation in this book can resolve some of the standard objections to free will. I also provide some suggested reading for those who would like to pursue this fascinating problem further.

In this chapter, we have seen how the transactional picture leads naturally to the 'arrow of time.' It is actualized transactions that generate the change which underlies all temporal concepts, and that change is measured by intervals of time. The generator of that change is the transfer of energy from one entity to another, and this is why we say that 'energy is the creator of time,' even though time is not really a substance. The actualized transaction is irreversible, since the offers and confirmations that operate 'behind the scenes' in Quantumland always do their work at the beginning of the process. This is because the exchange of offers and confirmations is a prerequisite for any actualized transaction, and all of those behind-the-scenes negotiations must be completed before any real energy is conveyed.[18] Trying to reverse this would result in energy being transferred without the necessary preliminary stages, which would not correspond to a real physical process.

---

[18]The term 'before' in this context describes the process of creating spacetime itself, so it is not the usual meaning of 'before' that refers simply to an earlier clock reading.

Another basis for the irreversibility is that offer waves are emitted from a single emitter, but they are generally confirmed by many absorbers. This one-to-many asymmetry between emission and absorption leads to a 'garden of forking paths,' which provides a natural basis for the inherent indeterminacy of future events, and an opening for genuine free will.

Chapter 9

'It from What?': Quantum Information, Computation, and Related Interpretations

'"It from bit" symbolizes the idea that every item of the physical world has at bottom [...] an immaterial source and explanation [...]'

John Archibald Wheeler[1]

In this chapter, we'll explore the concept of quantum information. We'll also take a brief look at some alternative interpretational approaches that attempt to explain quantum mysteries in terms of quantum information.

From 'Bit' to 'Qubit'

The classical bit vs. the quantum bit

The term 'bit' comes from the field of computation, and represents an answer to a 'yes/no' question. An answer to a question is just information; an abstract idea. So the quote from Wheeler that begins this chapter expresses the idea that the concrete physical world arises (somehow) from immaterial ideas. However, 'information' means different things to different people. Attempts to resolve quantum paradoxes in of terms of quantum information depend on defining what is really meant by that notion.

In the transactional picture, the 'it' of the ordinary classical realm of experience arises from a clearly-defined physical foundation, the offers and confirmations in Quantumland. These entities are more than mere information; they certainly contain information, but they exist in a more robust form than ideas or concepts in our minds, so they are

[1]Excerpted from Wheeler's essay "Information, Physics, Quantum: The Search for Links," *Complexity, Entropy and the Physics of Information,* edited by Wojciech H. Zurek (1990).

more substantial than the immaterial 'bit' notion referred to in Wheeler's quote. However, many approaches to quantum information treat information as something completely insubstantial. To examine these issues, we'll need a brief review of the idea of a 'bit' and how that is extended to the quantum version, called a 'qubit.'

A classical bit can take on only two values, **1** ('yes' or 'true') or **0** ('no' or 'false'). These two values are pieces of classical information. On the other hand, a quantum bit, or 'qubit,' can take on values corresponding to an arbitrary superposition of **1** and **0**, meaning that until the quantum state is measured, it could be either **1** or **0**. Thus, quantum information has a more 'slippery' character, corresponding to the ability of the two answers to be in a quantum superposition. Yet this slipperiness can lead to greatly expanded computational resources if it can be harnessed in practical applications. We'll consider some of those ideas later on in the chapter.

A qubit is made available in a real quantum computation by a real quantum system that can take on two possible values when measured, such an electron with a spin of either 'up' or 'down.' In the absence of a specific spin measurement, however, the electron is just an offer wave, which may be in a superposition of spin states; this is what makes it a qubit rather than a classical bit. In terms of our state symbols from Chapter 2, the two types of information look like Figure 9.1.

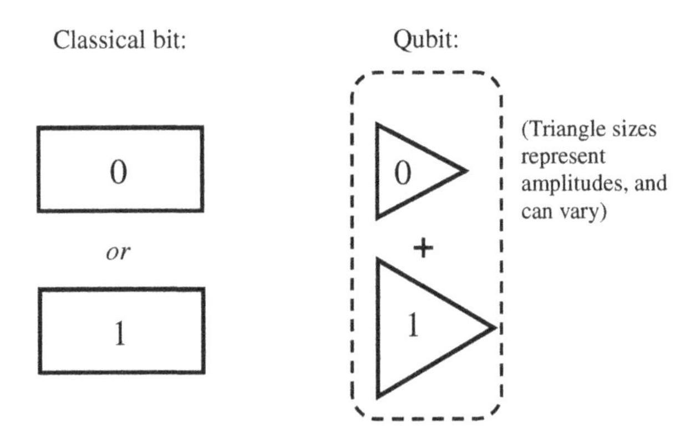

Figure 9.1. Contrast between a classical bit of information (left) and a quantum bit, or qubit, of information (right). For the qubit, the amplitudes of the zero and one states can continuously vary, within a constraint. In this illustration, the amplitude for | **0** > is smaller than that for | **1** >.

In the qubit illustration (Figure 9.1), the triangle sizes (quantum state amplitudes) can vary. However, they are subject to a constraint that arises from the Born Rule. Recall that the Born Rule tells us that we have to square the amplitudes to get the probability of an outcome. This leads to the requirement that the squares of the amplitudes have to add up to *1.0*. (A probability of *1.0*, or 100%, means that the outcome is certain to occur. Note that the probability of *1.0* is not to be confused with the label '**1**,' which is just the answer 'yes.') The requirement basically just says that when you do the measurement, you have to get an answer; in this case, either **0** or **1**.

In ordinary classical computation, we do not start with quantum systems, but rather with already-collapsed situations, such as states of a classical electromagnetic field.[2] So the big difference between the classical bit and the qubit is that the classical bit values are not in a superposition, and there is a concrete fact of the matter — an event (or set of events) within spacetime — as to which answer is the correct one. If this distinction seems hard to visualize, here's another way to picture it. The classical bit represents only two directions, 'up' and 'down,' while the qubit represents all possible directions from the center to the surface of a globe (in other words, all possible points on the globe). A few of those infinite numbers of possible directions are indicated on the right in Figure 9.2.

Even though the qubit seems to carry much more information than the classical bit, it turns out that much of that information is inaccessible; that

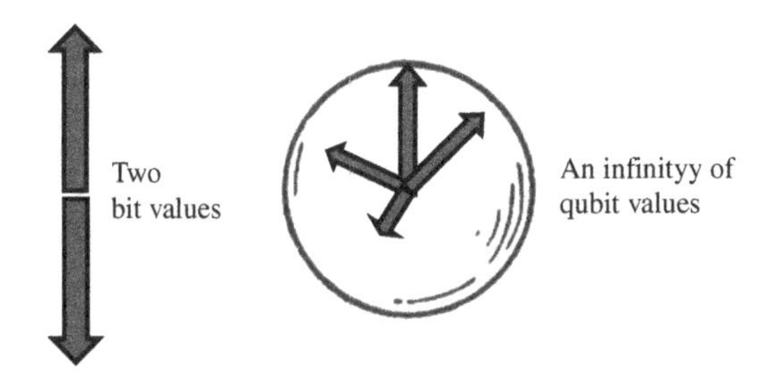

Figure 9.2. Left: a classical bit having only two possible values. Right: a quantum bit, having a continuum of possible values. The classical bit corresponds to only one particular axis through the sphere on the right.

[2]The classical electromagnetic field arises from very frequent and ongoing actualized transactions, and is thus the result of continual quantum collapses. This is described in detail in Kastner (2012), Chapter 6.

is, it cannot be retrieved directly at any arbitrary time in the computation process. The person who first showed this was Alexander Holevo, who produced proof to this effect (Holevo, 1973). This limitation on information retrieval from a qubit has become known as the 'Holevo Bound.' In the transactional picture, this is easy to understand: the qubit represents only possibilities (real, but sub-empirical), while any process of information retrieval must use detection to obtain an empirical phenomenon. This process, which is none other than measurement, always invokes an absorber response, and always results in 'collapse.'

We can picture the information retrieval process as the process of sticking a skewer through the center of the sphere in any orientation of our choice, which defines two possible outcomes at the points where the skewer pierces the sphere. Then, 'collapse' is the popping of the sphere to only one point on the surface, corresponding to the actualized outcome. An example of 'popping' to the outcome 'up' is illustrated in Figure 9.3.

The skewer represents the possible orientations of a magnet in a Stern–Gerlach (SG) experiment: the arrowhead represents the north pole of the magnet, and the arrow's tail represents the south pole. Recall from our discussion of the SG device that if a spin is not oriented exactly with the SG magnet, it is only a matter of probability (given by the Born Rule) as to whether the measurement will read up or down, according to how closely the spin is oriented relative to the SG magnet. We can choose any direction we wish, and measurement along the chosen direction will give

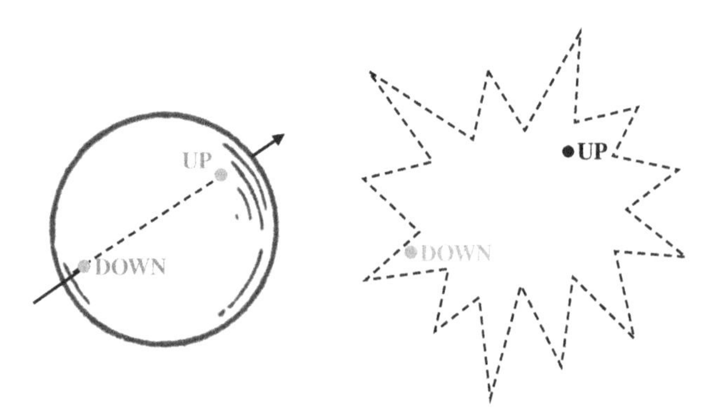

Figure 9.3. The sphere of possible values collapses to one of the two possibilities of 'up' or 'down,' upon measurement.

us two incipient transactions, one of which will be actualized. So, for example, suppose the actualized outcome is 'up.' That outcome is the one to which we can say 'yes,' while we must say 'no' to the unactualized outcome, 'down.'

This point may help to illuminate something initially puzzling about Figure 9.1: we only seem to have a superposition of the two possibilities: **0** and **1**. Where do all the other directions represented in the sphere come from? The full set of directions represents all the possible spin direction measurements that you could make; and they are all incompatible with one another. To get a 'yes' or 'no' answer, you have to pick a single direction, as illustrated with the skewer of Figure 9.3.

An example

As an example of the incompatibility of quantum measurements, consider again the SG spin measurement. The setting of the SG device specifies in which direction the spin of the electron will be measured. The direction chosen (represented by the magnet) can be along any diameter of the globe in Figure 9.4, where the outcomes 'up' and 'down' are indicated by U and D on the magnet.

Figure 9.4 shows an incoming well-defined quantum state of spin in the 'northeastward' direction (meaning we prepared it in that state, as discussed in Chapter 3). This corresponds to the value **1** — the answer 'yes' — for a measurement of 'northeastward spin.' But this state is in superposition with

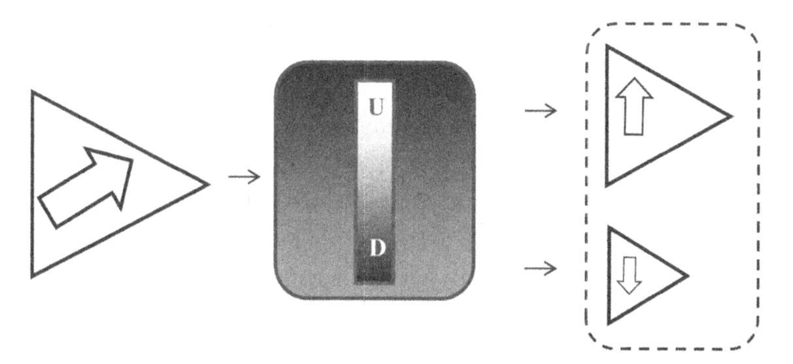

Figure 9.4. A 'northeastward' offer wave is separated into 'north' (up) and 'south' (down) offer waves by an SG device set to measure in the vertical direction.

respect to every other possible direction in which we could stick a skewer (as in Figure 9.4). This means that there is no 'yes' or 'no' answer as to whether its spin is along those other directions (even though there are tendencies described by the Born probabilities). This indeterminacy is shown in Figure 9.4 by the (uneven) superposition on the right for 'up' or 'down' with respect to the vertical direction. So, even though we start with a particular well-defined point on the sphere, corresponding to one skewer direction, the qubit is still in a superposition with respect to all the other skewer directions. This set of possibilities only gets 'shrunk down' upon measurement through the choice of a specific measurement direction, which dictates what outcome will correspond to 'yes' (**1**) and what outcome will correspond to 'no' (**0**). And of course, a crucial part of the 'shrinking' is the collapse due to the responses of absorbers (not shown in the figure).

So again, the basic message of quantum theory is that the set of physical possibilities attributable to a given quantum object is 'too big' to fit into spacetime. An offer has various tendencies for outcomes in any spin direction, but we cannot measure them all at once, and they are not actualized unless they are measured. Given a prepared northeastward possibility triangle as input into our experiment (as shown in the figure), we have to decide which direction to measure (i.e., what forces to impose on it, and absorber configuration to set up). Out of many possibilities, only one will be actualized. Although it has tendencies depending on the direction, it will be inherently unpredictable (unless we measure the northeastward direction, which will affirm that's what we started with).

No 'Cloning' of Qubits

Another aspect of the qubit that distinguishes it from the classical bit is that an unknown qubit cannot be copied. That is, you cannot set up a quantum computer as a 'copying machine' that will churn out copies of unknown input qubit states. As noted in Chapter 3, we can certainly prepare a state if we wish, but if we have not prepared it and we don't know what it is, we cannot simply copy it.

Let's look first at how an ordinary, classical computer can copy an unknown input. Suppose we want to copy an unknown classical bit, which can have a value of either **0** or **1** ('no' or 'yes'). Let's label our unknown bit '**?**,' since we don't know whether its value is **0** or **1**. We can make a 'copying gate' that will receive our unknown bit, along with a 'target' bit initially set to **0**. Think of the 'target' bit as a blank sheet of paper that can receive the information contained in the unknown bit. The copying gate reads the unknown bit '**?**.' It then leaves the target bit unchanged at **0** if the unknown bit has the value **0**, or flips it to **1** if the unknown bit has the value **1**. So we effectively get a copy machine. Let's represent the inputs by two numbers, the first being the value of the unknown bit and the second being the blank or 'target.' Then if **00** is the input, the output will be the same: **00**. If instead the input is **10**, the output will be **11**; the 'blank' has now been imprinted with the value '**1**.' The latter case is illustrated in Figure 9.5.

Now let's look at the quantum case. The big difference between this and the classical case is that we are dealing with a whole 'sphere' of possible values, and our 'copying gate' can work only with one axis from that sphere, which defines the answers 'yes' (**1**) and 'no' (**0**). Any input qubit that does not happen to align with that axis will be in a superposition, as illustrated in Figure 9.1 (right panel). But since the qubit is unknown, we cannot set our copying gate to the 'right' axis. We just have to pick an axis arbitrarily.

Suppose, just for the sake of argument, that we could copy a qubit, and we pick the north/south axis on the sphere for the orientation of our copying gate. When the gate saw an input corresponding to '**1**' it would copy that, and when it saw an input corresponding to '**0**' it would copy that (i.e., leave its target bit unchanged). But for a quantum system, both these cases can occur at the same time, since the qubit can be in a superposition of 'yes' and 'no.' This will be the case if the unknown qubit does not happen to be either 'yes' or 'no' with respect to the north/south axis, just as our

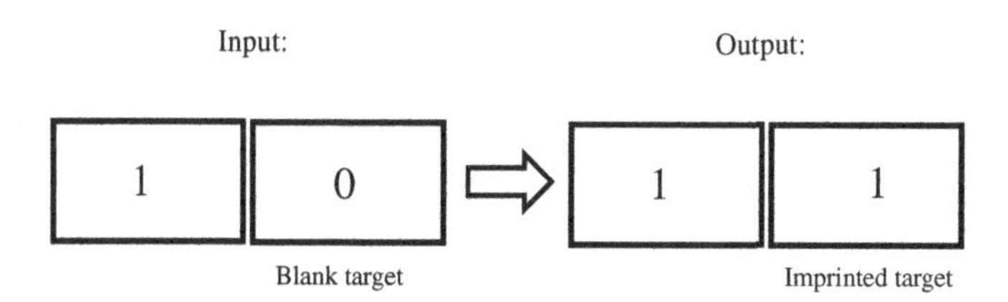

Figure 9.5. A classical bit is copied.

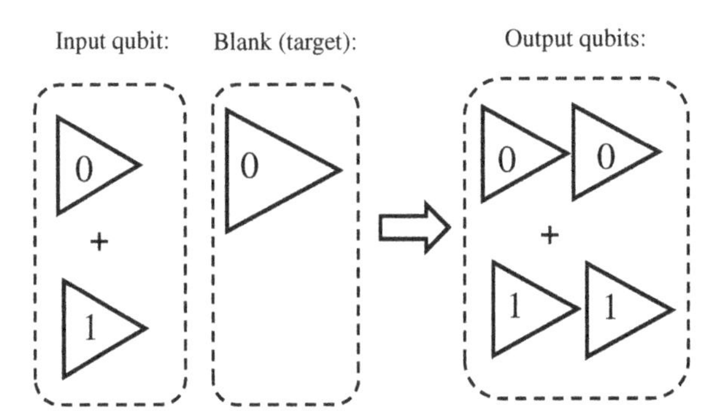

Figure 9.6. An attempt to copy an input qubit leaves both qubits in an entangled state, rather than providing a copy.

'northeastward' spin qubit ends up in a superposition of north/south in Figure 9.4.

Figure 9.6 shows us what this attempt at copying would look like. The 'blank' sees both the '**0**' and '**1**' inputs, and dutifully copies them both, thereby taking on the superposition of the original qubit! So the output that we get in this situation is an entangled state: a single superposition, with two qubits in each state of the superposition. The single superposition is indicated by the single dashed box surrounding both of the qubits on the right-hand side in Figure 9.6. This is the same kind of state as that of the two electrons in the Schrödinger's Kittens experiment.[3] In particular, it is definitely not a simple copy of the original qubit, which would have looked like the output in Figure 9.7. That is, if we could copy the qubit (which we cannot), the output would be two separate superpositions with each qubit in its own 'private' superposition. This is indicated in Figure 9.7 with a dashed box around each of the qubits. But quantum mechanics will not let us make this kind of individual copy of an unknown quantum. (We can of course copy a quantum if we know that it is either in state $|\mathbf{0}\rangle$ to $|\mathbf{1}\rangle$).

You might wonder: what about the entangled superposition of Figure 9.6? Is that some kind of copy that might be useful even if it's not the real

[3]The Schrödinger's Kittens entangled two-electron state was created in a very different way, however. There are many ways to create an entangled cat! One way to create a 'Schrödinger's Kittens' state is by radioactive decay in which two electrons are released from the nucleus at once.

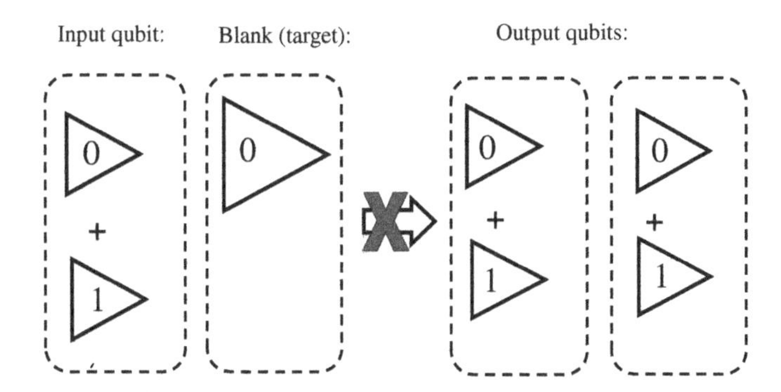

Figure 9.7. The right-hand side shows a simple copy of the input. However, quantum mechanics does not let us do this.

copy we wanted? The answer is, in general, no. This is because neither of the qubits even has a well-defined quantum state when they are entangled in this way, and we cannot get any useful information out of either of them by measuring them individually. Even the input qubit has lost its initial quantum state because of the interaction of the input qubit and the target qubit, so the attempt at copying has irretrievably altered our input! Thus, rather than gain any information about the input by copying it, we have lost whatever information we might have had to begin with.

In discussing the copying procedure above, we made use of a kind of operation that can be performed on a quantum system that changes its state, but does not collapse it. Specifically, we can change the state of a 'blank' qubit from $|\mathbf{0}\rangle$ to $|\mathbf{1}\rangle$ when it interacts with another qubit in the state $|\mathbf{1}\rangle$, but there is no collapse. This interaction, in which one can change a given initial quantum state to a different state gently, without collapsing it, is a kind of 'rotation.' That is, in mathematical terms, it amounts to pushing the arrow on the 'globe' pictured in Figure 9.2 around to a different direction (illustrated in Figure 9.8).

We actually discussed this in Chapter 4 in a different guise: it is the process in which a force acts on an offer wave to change it. To recall this process, see Figure 4.5, in which two electron offer waves act on each other (by exchanging virtual photons that convey the electromagnetic force), and are thereby changed. If we think of each of the electrons as a qubit, they are being 'rotated' to different states by their interaction, just as the qubit in the

Figure 9.8. A quantum state undergoes a smooth change, without collapse, to a different state. This is represented mathematically by a kind of rotation.

state $|\mathbf{0}\rangle$ can be rotated to the state $|\mathbf{1}\rangle$. No confirmation is generated at this stage, and that is why there is no collapse. We'll need this so-called unitary process to deal with further details of quantum computation.[4]

An Example of a Quantum Computation

If we can't use quantum computation to copy a qubit, what is it good for? Possibly quite a bit (pardon the pun!). It may allow us to greatly speed up certain computations. For example, finding the prime factors of a large number is a notoriously difficult problem for conventional computation. (Prime numbers cannot be divided by any number except themselves and 1 and still yield a whole number answer. The prime factors of 6 are 2 and 3 ($2 \times 3 = 6$).) Prime factorization on a classical computer is extremely time-consuming. This is especially so if the prime factors are very large, such as 100 or more digits long. These sorts of very large prime numbers are used in computer security codes.

Working with qubits instead of classical bits offers the possibility of a significant computational speedup, due to the ability of the qubit to be in a superposition of an answer to a 'yes/no' question. The superposition appears to provide a way to perform many computations simultaneously instead of having to do them one by one: in the case of a single qubit, this

[4]The term 'unitary' is based on the word 'unity' and means that the possibility triangle keeps its size intact. In contrast, in a collapse process, in general the possibility triangle shrinks. This corresponds to the vanishing of the possibilities not actualized in a transaction. The collapse process is therefore called *nonunitary*.

would amount to having as inputs both 'yes' and 'no,' and processing both of them at the same time. This possibility is referred to among quantum computation researchers as 'quantum speedup'.

One of the pioneers of quantum computation is mathematician Peter Shor. He developed a quantum scheme for performing prime factorization. Its key feature can be viewed as a 'rotation,' as discussed in the previous section. But Shor's rotation is performed on a higher-dimensional 'globe' than the one pictured above. This larger 'globe' is the space of quantum possibilities available to not just one qubit but to a collection of them. The larger the number to be factored, the more qubits are needed, and the larger the space of possibilities that applies to them. You can think of this larger globe as a multi-dimensional globe, just as a cube can be seen as a multi-dimensional version of a square. Thus the Shor rotation is a kind of mega-rotation; a rotation 'on steroids,' if you will. A more appropriate term for this complicated process is a 'transformation'.

The Shor transformation acts like a kind of sieve that filters out all the wrong answers, and lets only the right one through. It is performed by doing a collection of simpler rotations that place the input state, which contains information about the number to be factored, into a complicated superposition of the basic qubit 'yes/no' states. The different states that make up the superposition interfere with each other (just as we get interference in the two-slit experiment). The interference appears because of some esoteric features of number theory that we can't go into here, but the key point is that all the states with the wrong kind of numerical property end up cancelling each other out. Meanwhile, the one state with the right kind of property gets reinforced, so that its amplitude becomes very large. (This is the same sort of process that produces the bright stripes in an interference pattern, as discussed in Chapter 2, except in this case there is a single 'bright spot' and everything else is dark.) The result is that when a collapsing measurement is finally performed, ideally only that one particular outcome gets through the sieve, and that outcome provides a key piece of information leading to the correct factorization.

In the transactional picture, the Shor factorization process amounts to the following: an initial physical possibility, $|\,\mathbf{N}>$, containing information (from number theory) about the number N to be factored, undergoes a series of interactions that do not involve absorber responses. (This is similar to the way in which an offer wave goes through both slits of a two-slit experiment and does not get a confirmation from either slit.) These interactions transmute the initial possibility offer into a different one, | **R** >. (This is a kind of 'rotation' of the arrow representing the collective, entangled quantum state of all the qubits involved; a unitary process.) This new possibility is then allowed access to a set of absorbers; you can think of these as detectors like those in an SG experiment. But it will only generate a confirmation from one absorber, because (ideally) it will not reach any of the others. This is the same thing that happens in an SG device with the magnetic field set for the same direction as the spin of an incoming electron 'offer': the only absorber response it will receive is from the one accessible to it. Thus, we have the kind of situation depicted in Chapter 3, reproduced in Figure 9.9 but labeled 'R' in accordance with the desired information.

Thus, at least ideally, we have only one possible transaction. The offer waves for the wrong answers cancel out, and the offer wave for the right one is reinforced. In the ideal case, this outcome is actualized with certainty, and it contains the information we are looking for! (There are other detectors for other possible integer values, but these cannot respond to the wrong offer possibility, just as the 'down' detector of the SG device cannot respond to the 'up' offer.) Of course, in real life, there will be other incipient transactions, but with much smaller probabilities. The smaller those probabilities can be made, the smaller the chance of error in the

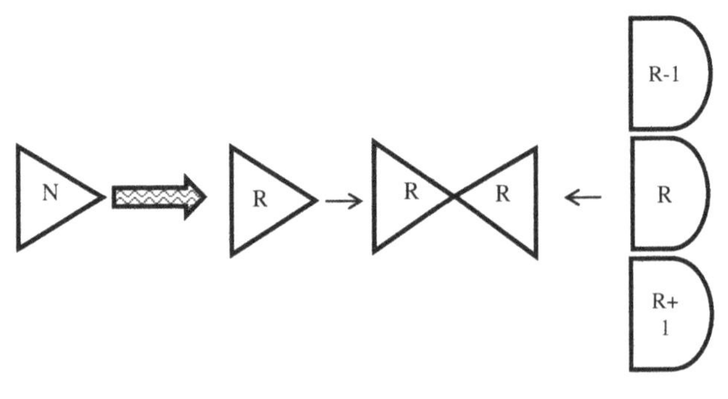

Figure 9.9. The Shor transformation in the ideal case.

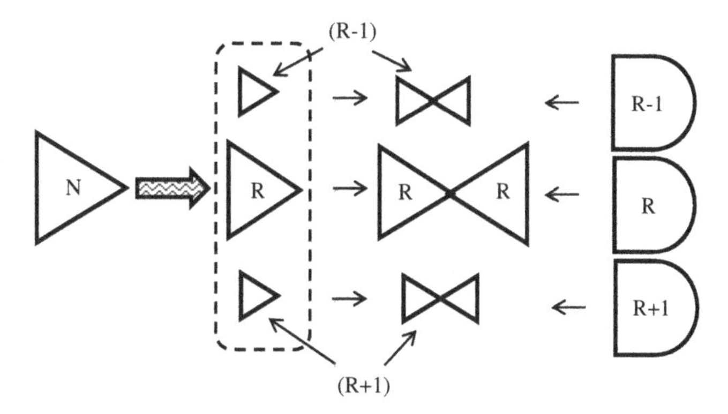

Figure 9.10. The Shor transformation in the non-ideal case. There is a small probability of getting an outcome corresponding to a wrong answer.

computation. The non-ideal case is shown in Figure 9.10. Thus, the incipient transaction corresponding to the desired value has a much larger probability than the others. The result is that the algorithm produces the right answer most of the time, but not all of the time.

As you can see, the Shor transformation is an ingenious idea. The tricky part is implementing it in the real world, since it relies on preserving a delicate superposition of quantum states. As of this writing, the process has been successfully implemented only for factoring small numbers such as 15 and 21, for which we already know the prime factors! It remains to be seen whether technology can be developed to successfully factor larger numbers.

The idea of a 'quantum speedup' — which makes use of the larger mathematical space describing the quantum possibilities — is often visualized in terms of the 'Many Worlds' interpretation of quantum mechanics. Indeed, another quantum computing pioneer, David Deutsch, has advocated the idea that the computation is going on in many 'parallel universes' at once. But it's not at all necessary to adopt the many-worlds view in order to understand these processes: they can be seen quite naturally as interactions among the many physical possibilities in a transactional picture. Upon measurement, only one outcome occurs, and this is the actualized transaction. And again, this is why much of the information being tossed around during the quantum computational process is subject to the Holevo bound; i.e., it is inaccessible, because all this information lives in a sub-empirical realm of possibilities and cannot be accessed empirically, except through a collapsing measurement.

The preceding is just an overview of what kinds of processes are involved in quantum computing, and how their interpretation differs in the transactional picture, which takes absorption into account as a physical process. The usual account is missing the backwards triangle (confirmation) that initiates an otherwise mysterious collapse of the quantum state. Readers interested in more details on the Shor transformation can find many presentations on the internet.[5]

Information and Interpretation

Thus far, we've examined the concept of information in its classical and quantum aspects. We've seen how quantum information, as contained in the 'qubit,' holds promise for making computation more powerful, even though the feasibility of quantum computation remains uncertain. Computing is basically the manipulation of information as embodied in various types of physical systems. The root of 'information' is the word 'form': having structure, but immaterial. This brings us back to the quote of John Wheeler that began this chapter, and the question of whether 'It,' meaning the physical world of experience, comes from 'Bit,' meaning information. This is a question concerning what it is that quantum theory is describing. We return here to that interpretive question.

As noted at the beginning of this chapter, some researchers have proposed that quantum theory can be understood as a theory about 'quantum information.' Ultimately, though, appealing to the concept of information provides no escape from the conceptual challenges that quantum theory presents to us, because we are still faced with the basic choice of realism or anti-realism about quantum theory. Realism about quantum theory implies that quantum states describe real entities that exist in the world. Since these entities cannot fit into the phenomenal spacetime realm, a realist approach requires us to enlarge our world view to one that encompasses a larger

[5]One presentation with all the steps of the algorithm is by Dhushara: http://www.dhushara.com/book/quantcos/qcompu/shor/s.htm.

realm of real physical possibilities.[6] This is the approach that has been advocated in this book. On the other hand, antirealism about quantum theory boils down to the claim that quantum states describe only our knowledge. In that approach, the 'quantum information' is about our subjective experiences, and not about anything independently existing in the world.

However, there is a rather popular 'middle way' between realist and antirealist approaches, which we will now consider. That approach assumes that quantum states (our possibility triangles as presented in this book) are just descriptions of our imprecise or approximate knowledge about a real property of the quantum system that is hidden and inaccessible to us. Thus, this approach has an aspect of antirealism — taking quantum states as primarily about subjective knowledge — but it also has an aspect of realism: the idea that there is some real property, or state of being, of an independently-existing entity out there in the world. In what follows, we'll see first why this approach has seemed promising, but we'll also find that it cannot work.

Let us call this approach the 'shell game,' because it is similar to the game in which a magician hides a pea under one of several shells (Figure 9.11). The pea represents the real property of the system, and the shells are various quantum states that could be assigned to it based on our experimental procedures and measurements. The idea is that as new results of measurements

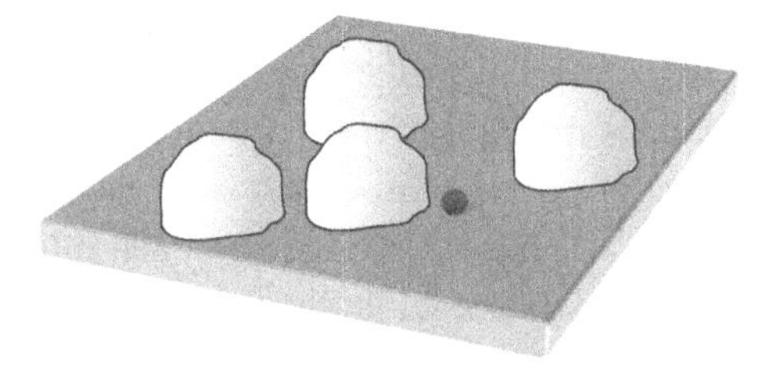

Figure 9.11. Is quantum theory a shell game? A recent theorem says 'no.'

[6]There are other realist views, such as the Many Worlds approach or the Bohmian theory, which also must radically enlarge our view of what exists. I have argued that the Many Worlds Interpretation fails because the worlds cannot really split, at least not via 'decoherence' as is generally assumed (Kastner 2014c). See also my blog post at http://transactionalinterpretation.org/2014/07/16/why-the-world-cannot-really-split-in-the-many-worlds-interpretation/.

become available to observers, their subjective knowledge increases, and therefore their ignorance decreases. In terms of the shell game, it is like picking up one shell, finding it empty, and then throwing that one away. Now we know the pea must be under one of the fewer remaining shells. That reduction in our ignorance could then be what is meant by the 'collapse of the wave function': it is just a reduction in mental uncertainty.

Here's an example in which such a 'shell game' interpretation of a physical situation might make good sense. Suppose you need a home repair item, say a new handle for a cabinet, and you decide to go to the new hardware store in town to get one. We could label your initial state of knowledge about the handle's location with the symbol $| \mathbf{S} >$, representing the entire 100,000-square-foot store. Now you make your first 'measurement': you walk into the store and see a sign, 'Hardware,' hanging at the far right of the store. As you approach, you see that the hardware section occupies about ¼ of the store area. Your state of knowledge of the item's location has just 'collapsed' to $| \mathbf{H} >$, which represents a specific 25,000-square-foot region. Now you make another 'measurement': you ask an employee where the cabinet handles are, and he points you to Aisle 12. Once again, your state of knowledge 'collapses' to the more focused, precise state of $| \mathbf{A} >$, which represents only that aisle. Finally, as you walk down the aisle, you spot the handle you are looking for. Your knowledge state finally 'collapses' to $| \mathbf{X} >$: the precise spot where the handle is sitting on a shelf. Each of these collapses, to S, H, A, and X, pertain strictly to your subjective knowledge; none of them is an objective, physical collapse.

Taking the collapse as mental provides a tempting way to relieve ourselves of the challenge of understanding nonlocal wave function collapse as a real physical process.[7] Thus, the above approach might seem to allow us to preserve a traditional, classical understanding of reality. The idea is that these 'state collapses' were about your knowledge, not about the handle; it had its position all along, you just didn't know it. There is no quantum indeterminacy or instantaneous collapse that seems to violate relativity (in which the effects of the collapse would appear to propagate at infinite speed). So using a 'Shell Game Interpretation' of the quantum

[7]If we expect collapse to be a spacetime phenomenon, it violates relativity by apparently propagating at infinite speed, and by picking out a preferred frame.

state — just as in the above hardware store situation — might seem like a good alternative to taking 'collapse' as a real physical process.

However, it turns out that quantum mechanics itself will not allow us this interpretation of the quantum states, as appealing as it may seem to some. When we work out the details, the Shell Game approach turns out to conflict with the well-corroborated predictions of quantum theory. This was shown by the team of Pusey, Barrett, and Rudolph (2012). For the interested reader, a simplified version of the proof is presented in Appendix C. Basically, the proof shows that quantum mechanics allows us to replace all the white shells with a different set of shells (say all black), such that the pea cannot be found under any of them. But the 'Shell Game Interpretation' does not allow this: it says that the pea must still be there. So we have a contradiction, and since quantum theory has already survived many experimental tests, the 'shell game' approach cannot be right.

Information about What?

As noted in the Preface, the central interpretational problem of quantum theory is to answer the difficult question 'What is quantum theory really about?' The question is difficult because the things that it was developed to describe, like electrons and other subatomic particles, behave so strangely that it seems that they may not be entirely real (at least, according to the usual, classical standards of 'reality'). For example, these quantum objects don't obey the cosmic speed limit of relativity when it comes to correlations (as in Schrödinger's Kittens), and they seem to 'collapse' in a way that also seems to defy that limit; the effects of collapse seem to propagate at infinite speeds. They also seem to give us answers that depend in a crucial way on how we ask the questions. This situation has led some researchers to say that quantum theory is simply about 'information,' and that's the end of the story. While some of those 'information-only' approaches don't deny that there might be some sort of reality out there, they claim that the theory itself is primarily about the knowledge and subjective experiences of observers, rather than about that reality.

One reason that some researchers have given up on realist explanations is that they assume that if something cannot exist within spacetime it does not exist, period. So from that standpoint, quantum entities don't exist. But yet somehow they do, because they have apparently given rise to the theory that describes what sorts of phenomena we will experience when we conduct experiments on them. If one does not allow for the existence of real entities that transcend spacetime, one becomes tied into logical and linguistic knots in which quantum theory is taken as being 'about information', but the information is tacitly assumed not to be about anything that really exists. Alternatively, the information is taken to be about our interactions with something held to be wholly inaccessible; not just to sensory perception, but even to any kind of rational inquiry.[8]

A far more straightforward interpretation is that quantum theory is about new kinds of entities that are not confined to the spacetime realm; they are sub-empirical. That way, it's very simple: they contain information because they are real. Quantum information is about real quantum systems, just as Boltzmann's successful theory was about real things: atoms. Thus, the interpretation presented in this book, called the Possibilist Transactional Interpretation (PTI) in the scientific literature, suggests that quantum theory works so well simply because it is describing entities that underlie the phenomenal world of experience. This is so even if those entities behave in strange and unexpected ways, such as the past-directed responses of absorbers. Indeed, it is that strange absorber response that helps us make sense of the previously mysterious Born Rule.

This approach fits naturally within the history of scientific advance, in which the success of theories has forced us to alter and/or expand our world view. For example, Einstein's theory of relativity taught us that space and time were intermingled, and that they were not absolute and immutable containers for events, as Newton had thought. That was a revolutionary message about reality; one which was ultimately embraced. We face a similar situation today: quantum theory gives us an unexpected message about reality. We should not retreat from that message by saying that the theory is not about anything physically real or rationally intelligible, just as the scientists of the 20th century did not retreat from the

[8]This is reminiscent of the way in which Kant provided no explanation for how his noumenal entities interacted with the knower to produce the phenomenal world.

message of relativity by saying that it's so strange that it must not describe anything physically real or rationally intelligible.

In this chapter, we've discussed quantum computing and quantum information, and explored the extent to which the idea of 'information' can help to answer quantum riddles. We've seen that taking quantum theory as being about 'information' begs the question of what that information is about. In the end, we have to decide whether the information is about something physically real or if it is about subjective impressions and experiences. In the next and final chapter, we'll investigate how the choice of a realist option provides us with an exciting opportunity to expand our view of reality, just as previous scientific revolutions have always done.

Chapter 10

Epilogue: The Next Scientific Revolution

'Man cannot discover new oceans unless he has the courage to lose sight of the shore.'

Andre Gide, French novelist

Science is the endeavor of gaining knowledge about reality. Its vital tools are: (1) empirical observation and (2) logical and mathematical reasoning in the form of theories. An important 'quality control' of scientific theorizing is that it must be supported by empirical observation. Observation alone is the mere cataloging of phenomena without explaining them, and theorizing alone is speculative 'armchair philosophy' that doesn't necessarily explain anything about the empirical world. So physical science needs both tools (1) and (2) in order to provide effective explanations of physical reality.

Scientific advance has always been a process requiring its practitioners to relinquish firmly-held views about the nature of reality; views based on incomplete or limited types of observations (tool (1)). This is because methods of observation are always improving. Theories accounting for less-accurate observations often run into trouble in the face of better ones, so those theories must be improved, or even scrapped entirely.

Here's an example that is familiar to everyone: thousands of years ago, it was assumed that the Earth was flat, based on empirical observation. That is, a casual glance at the terrain around us with the naked eye yields the distinct impression of a flat plane connecting to the sky at a more or less straight horizon (we disregard local imperfections such as hills or valleys). We believe what we see, and what we see in such an observation is that 'the earth is flat.' This was the understanding of physical reality up until the time of the ancient Greeks. At that point, things began to change. Observations became more sophisticated, particularly in the context of sea travel, which revealed land sightings 'beyond' the horizon as one ascended in height on a ship's mast or on an ocean swell. The Greeks, to their credit,

Figure 10.1. The 'Flammarion,' a famous depiction of the flat Earth cosmology, by an anonymous artist.

realized that a flat-Earth picture could not accommodate these observations. The discovery of a spherical Earth could be considered the first big 'scientific revolution.'

However, the 'flat Earth' belief still held sway among many societies long after the cosmological advances of the ancient Greeks (indeed, the Flat Earth Society's website is actively maintained to this day). This fact is testament to how difficult it is to shake off the 'obvious' impressions of our senses. Figure 10.1 shows a famous engraving of unknown origin, first appearing in Camille Flammarion's *L'Atmosphere* of 1888. The image depicts the flat-Earth conception, under the inverted celestial bowl of the 'firmament.' Flammarion wrote of this image:

> Our ancestors imagined that this blue vault was really what the eye would lead them to believe it to be; but, as Voltaire remarks, this is about as reasonable as if a silk-worm took his web for the limits of the universe.

Thus the quest for knowledge of reality has continually demanded that we expand our conceptual toolbox concerning the boundaries of what is assumed to be 'real.'

The second scientific revolution occurred during the Renaissance, when the idea of the Earth as the center of the universe had to be relinquished in favor of the sun-centered theory proposed first by Nicolaus Copernicus and later refined by Johannes Kepler. It was 'common sense,' based on 'direct observation,' that the Sun, Moon, and all other celestial bodies were moving around the Earth, while the Earth itself remained stationary. Indeed, we still see the same phenomena today! It was only the increasingly-sophisticated observations of the heavens (some of these by Copernicus himself, and later by Tycho Brahe) that began to cast doubt on this 'obvious' theory concerning the configuration and motions of the celestial bodies. The doubt arose from observations of retrograde motion on the part of some of the planets: at certain times of the year, they seemed to halt and then move backward in the opposite direction. Ptolemy attempted to account for these anomalous motions in the Earth-centered theory by adding additional small circular orbits to those planets (these he called 'epicycles'[1]). However, as the observations became more sophisticated and detailed, more and more 'epicycles' became necessary, and the theory lost its initial simplicity and elegance as it became more and more encumbered by these ad hoc alterations.

It was in the context of these new observations that Copernicus first proposed his Sun-centered theory, which neatly accounted for the observed retrograde motions in terms of the relative motions of the Earth and other planets around the Sun. But (as is well known) the Sun-centered ('heliocentric') theory was fiercely opposed by many; and not just on the ideological/religious grounds of the Church of that era. Resistance to the heliocentric model on the part of ordinary people of the time is quite easy to understand: even in our day, we 'see' the Sun and other celestial bodies circling the Earth, and we 'feel' that the Earth is not moving! In order to accept the heliocentric theory, we must disregard the immediate evidence of our senses in favor of an abstract construct that more elegantly accounts for sophisticated astronomical observations (e.g., retrograde motion) that we don't see unless we undertake painstaking, long-term studies of the

[1]Actually, Copernicus' model retained some epicycles because he assumed the planetary orbits were circles. It was Johannes Kepler who first arrived at the mathematically-correct model, which generalized the circles to ellipses, thereby eliminating the need for ad hoc epicycles.

heavens. The more successful heliocentric theory in many ways seems to deny our empirical experience, yet we all accept it now, although its introduction was highly controversial and its ultimate acceptance was revolutionary.

The third scientific revolution, the quantum revolution, began around 1900 with Max Planck, Werner Heisenberg, Niels Bohr, Erwin Schrödinger, and Albert Einstein. Yet, more than a century later, the revolution is not over. Controversy still rages, and there is a marked lack of consensus among physicists and philosophers concerning the implications of the theory for our understanding of physical reality. Let's consider this situation against the backdrop of the previous two revolutions. These earlier revolutions involved the relinquishment of two firmly-held beliefs about reality.

1. The earth was thought to be flat.
2. The earth was thought to be the center of the universe.

I suggest that the reason this third, quantum revolution is not yet complete is because there is another very firmly-held, apparently 'common sense' view about physical reality that needs to be relinquished:

3. Spacetime is thought to encompass all of reality.

Recalling our iceberg of Chapter 1, the 'tip of the iceberg' represents the spacetime realm. View 3 asserts that reality only consists of the tip of the iceberg, the part that can be seen above the water. But since the quantum objects described by the theory have a structure that is 'too big' for them to fit into spacetime, they must live beyond it, in a mathematically larger[2] realm of possibility (depicted as the portion of the iceberg below the surface). Quantum objects are therefore hidden from direct observation.

Specifically, quantum theory describes quantum states as multi-dimensional[3] as well as complex. A common response to this fact by those

[2]That is, having many more dimensions.

[3]That is, states applying to N particles are 3N-dimensional. So, for example, the quantum state applying to a Schrödinger's Kittens electron pair is six-dimensional, in addition to its complexity.

who adhere to view 3 is to assume that quantum states do not refer to something real; that they must represent some sort of 'information.' However, taking quantum states as describing 'information' begs the question: information about what? If the information is taken as only mental, the approach is antirealist; i.e., it denies that quantum objects really exist.[4] If we think that quantum objects such as photons, electrons, and atoms really do exist, the simplest explanation for the strange multi-dimensionality of the theory that describes them is that spacetime is not all there is to physical reality.

Taking quantum objects as physically real, but existing in a larger realm beyond spacetime, is a challenging conceptual leap. So it's not surprising that it is taking some time for the quantum revolution to reach its fulfill-ment. It should be kept in mind that it took roughly two centuries for the heliocentric theory to be fully accepted, so perhaps several centuries is a standard length of time required for these 'paradigm changes' (as physics historian Thomas Kuhn called them (Kuhn, 1962)).

Moreover, there is an aspect of this conceptual transformation that goes beyond the earlier ones. We can obtain specific empirical observa-tions that confirm that the Earth is not flat, and we can also obtain spe-cific empirical observations that confirm that the Earth goes around the Sun rather than vice versa. But, by definition, there is no empirical observation that can confirm for us that there is more to physical reality than what is in spacetime, because spacetime is the realm of observation! We cannot directly confirm via empirical observation that there is any-thing beyond the 'tip of the iceberg.' The leap we make must be a con-ceptual one, on the basis that it is the best explanation for what we can see, especially in view of previous scientific episodes suggesting that observed phenomena never seem to be the whole story about what is 'really going on.'

---

[4]Alternatively, some researchers who wish to retain view 3 are attempting to get around the Pusey, Barrett, and Rudolph (PBR) theorem (discussed in Appendix C) by using dif-ferent kinds of elements of reality. This relegates the quantum formalism to a measure of the ignorance of observers, rather than a structure that reflects reality as it is. It also requires the elements of reality to be specially tailored so as to escape the inconsistency revealed by the PBR theorem, which introduces new complications not present in quantum theory itself.

Accepting the idea that what we can directly observe is not the whole picture involves a kind of humility: the acknowledgement that our senses can deceive us. Descartes (1641) pointed this out, and he preceded that with a prophetic argument that it is a mistake to think that 'there is nothing more to reality than what we can touch' (1633, Chapter 4). The quantum revolution can be seen as a logical continuation of the pattern set forth in the preceding two revolutions: the pattern is one in which we must acknowledge that our immediate observations are not sufficiently wide in scope to be informing us of the 'big picture.' Our point of view is revealed as too local, too partial, and in need of generalization. The idea that needs generalizing in the quantum revolution is our notion of what is 'real'; spacetime is not the whole story.

Expanding our view of what is real allows us to see a further defect in Niels Bohr's approach to interpreting quantum theory. Bohr famously said that 'It is wrong to think that the task of physics is to find out how Nature is. Physics concerns what we can say about Nature.' (Petersen, 1963). But there is an implicit unsupported assumption in this statement; namely, that 'One cannot say how nature is.' Yet quantum theory does describe 'how nature is' at subtle levels, so certainly the theory is doing just that. Of course, one has a hard time saying 'how nature is' if one wrongly assumes that all of nature is contained in spacetime, which is what Bohr was implicitly doing.

Quantum theory is indeed telling us 'how nature is.' All we need to do is to let go of the idea that knowledge must be limited to the spacetime realm and its concrete actualities. We can talk about possibilities, can't we? So we can indeed say something about nature, if this proposed interpretation is correct. And what we can say is this: nature encompasses a vast realm of unactualized possibilities that give birth to the actualized, physical world of spacetime. Quantum theory is the theory that describes those possibilities and certain aspects of their behavior, although it does not provide a deterministic account, since the possibilities are not deterministic entities. After all, it is the nature of possibilities to 'keep their options open'!

Thus, the picture of reality proposed in this book contains new elements — physical possibilities — that need to be added to our world view in order to understand why quantum theory successfully accounts for the phenomena observed in experiments with microscopic objects such as atoms. These physical possibilities are entities previously unsuspected, just as the lower portion of the iceberg is not suspected (until our ship runs aground on it!). Indeed, there are many historical examples of apparently abstract mathematical inventions and calculations turning out to contain real physical content that needs to be included in our world view. One such abstract notion was the invention of the 'imaginary' number *i*, defined as the square root of −1 (which has no real square root); we discussed these numbers in Chapter 2.[5] Recall also Freeman Dyson's remark (2009) that:

> [T]he discoverers of the system of complex numbers thought of [it] as an artificial construction [...] a useful and elegant abstraction from real life. It never entered their heads that this artificial number system that they had invented was in fact the ground on which atoms move. They never imagined that nature had got there first.

Similarly, quantum theory is not just a 'useful and elegant abstraction,' a computational tool that (somehow) enables us to predict phenomena. It works because it describes something real. And this is by far the simplest explanation for why it works so well.

One way of evading the difficult interpretational questions we've considered in this book is to say that they are 'meaningless' or that we should accept that nature will never provide an answer. But this is just an admission of defeat;[6] one could say the same thing in response to any of the three riddles at the beginning of Chapter 2, if one did not happen to know the correct answer. The correct answer to a difficult question is always something that demands a revision of one's preconceptions. Now, one could certainly devise a riddle that didn't have an appropriate answer, but nature really does act according to quantum theory, so nature is clearly doing something, whether we happen to understand it or not. This distinguishes quantum theory from an arbitrary riddle which could be specifically designed to be unanswerable.

---

[5]Another historical example in which an apparently abstract calculation led to a mathematical object that turned out to describe a physically real process is in the discovery of electron spin (a physical property discussed frequently throughout this book).

[6]In fact it is worse than an admission of defeat. It is a portrayal of failure as success.

But we should also note that even a riddle designed to be unanswerable could turn out to be answerable through sufficiently creative thought. Here's an example: 'What is the square root of −1'? As noted above, mathematicians 'invented' an answer to this riddle by 'creating' the imaginary number *i* and the complex numbers. As Dyson (2009) noted, these 'inventions' turned out to be describing reality. Thus, as Jeeva Anandan (1997) observed (in Chapter 1), nature always seems to be way ahead of us in richness of imagination.

Finally, I should note that it has been argued by some philosophers that adding new elements to our word view is to be avoided, based on an argument by William of Occam, known as 'Occam's Razor.' The argument basically states that when one is considering competing hypotheses, one should choose the simplest one. However, Occam's Razor is not really an argument for the simplicity of a world view, but rather for the simplicity of *explanation*. The simplest, most straightforward explanation for why quantum theory works is to say that its mathematical objects (such as quantum states, the possibility triangles in this text) represent physically real things; that they really are acting, even if 'behind the scenes' of the spacetime theater, to create the phenomena that we do see.

Just as researchers of the past have discovered new frontiers — first beyond the surface of the Earth and later beyond the Milky Way Galaxy — quantum theory has now led us to the next frontier. All we need to do is recognize what the theory is disclosing to us: a vast domain of physical possibility that lies hidden beneath the tip of the spacetime iceberg. There is more to reality than meets the eye.

Appendix A

How Absorption Illuminates the Measurement Theory of John von Neumann

We noted in Chapter 4 that standard approaches to quantum theory do not provide a clear distinction between offers, incipient transactions, and actualized transactions. This is because they do not take absorption into account. Recall that it is absorption that gives rise to one or more 'bracs,' < **p** |, which trigger the formation of one or more incipient transactions represented by 'bow ties.' Since standard approaches do not include absorption, they can give no physical account of this measurement process. Nevertheless, they still make use of the 'bow ties' to obtain predictions from the theory; these are called 'projection operators,' and they represent observable outcomes. We talked about this in Chapter 3, but in this Appendix, we'll make it more precise, with reference to some important work by the mathematician John von Neumann.

It was von Neumann who, in the early 1930s, developed a rigorous mathematical framework for quantum theory. That formulation included specific mathematical rules that seemed to apply to any measurement, though von Neumann could not explain what constituted a measurement in physical terms. Von Neumann's rule consists of two parts: (1) a quantum state transforms into a set of projection operators, each multiplied by a squared amplitude; and (2) this set of projection operators collapses to only one projection operator, in which the probability of that outcome is given by the squared amplitude that multiplied it. Previously in this book, we described amplitudes as the 'sizes' of the possibility triangles, but in the actual theory, an amplitude is a complex number that is given by the inner product of two quantum states (represented in this book by triangles). In terms of our 'bracket' notation, it is a bracket. It looks like this: <X|Y>. It can be roughly understood as the amount that the two states

$|\,\mathbf{X} >$ and $|\,\mathbf{Y} >$ overlap. The reverse order, $<Y|X>$, is the complex conjugate of $<X|Y>$. The absolute square of the amplitude $<X|Y>$ is written in condensed form as $|<X|Y>|^2$ and is obtained by multiplying $<X|Y>$ by its complex conjugate: $|<X|Y>|^2 = <X|Y><Y|X>$. (This squared *inner product* is a probability; the one that appears in the Born Rule. In contrast, a 'bow tie,' representing a projection operator, corresponds to what is called the *outer product*.)

For clarity, consider a specific example: a light source emitting a photon offer wave in a superposition of four different momenta, where that superposition is described by $|\,\mathbf{A} >$, and a set of absorbers corresponding to each of those component momenta. This means we are measuring momentum. Suppose the component momenta are numbered 1, 2, 3, 4. In that case, von Neumann's rule says that first, $|\,\mathbf{A} >$ transforms into a set that looks like:

$$|<1|A>|^2\ |1><1| + |<2|A>|^2\ |2><2| + |<3|A>|^2\ |3><3| + |<4|A>|^2\ |4><4|.$$

Each of the 'bow ties' (projection operators) corresponds to a particular momentum outcome. Only one of these outcomes is found each time, with a probability corresponding to the squared amplitude that multiples it. For example, we would find the result '3' with the probability $|<3|A>|^2$. This is von Neumann's rule, and it works because the absorbers generate confirmations described by $< \mathbf{1}\,|$, $< \mathbf{2}\,|$, $< \mathbf{3}\,|$, and $< \mathbf{4}\,|$, even though that was not part of his formulation. He simply noted that this is the correct mathematical formulation that seems to describe whatever it is that goes on in a measurement.

Thus, even though standard approaches to quantum theory don't physically distinguish between them, von Neumann's rule makes reference to the following three distinct objects: offer waves $|\,\mathbf{p} >$, incipient transactions $|p><p|$, and actualized transactions $\boxed{\mathrm{P}}$. There hasn't been general recognition of these three objects as physically different entities because they have been treated merely as part of a mathematical recipe that works, even though (generally) nobody knows why. So standard approaches only treat the quantum states $|\,\mathbf{p} >$ as representing quantum objects; they have no name for the physical entities corresponding to $|p><p|$ and $\boxed{\mathrm{P}}$ because these are not part of the standard approach. This leads to confusion, since

the same word, for example 'photon,' is used for an offer wave, $|\,\mathbf{p}\,>$, and an actualized transaction, $\boxed{\mathrm{p}}$. Standard ways of applying and interpreting quantum theory cannot distinguish between these entities, even though they are part of the mathematical formalism.

When a theory contains a mathematical calculation that gives the right answer but the physical reason for it is unclear or thought to be nonexistent, the calculation (and sometimes the entire theory) is merely an instrument for predicting phenomena rather than an explanation for the phenomena. This has been the situation in quantum theory for nearly a century. But if we include absorption as a real physical process, we readily see that quantum theory, including von Neumann's measurement calculation, is not just an instrument. It describes real, physical processes, albeit with some surprising features. The surprise should be welcomed as the unearthing by quantum theory of subtle aspects of reality that could not be seen with the limited tools of classical physics.

Appendix B

Free Will and the Land of the Quantum Dominoes

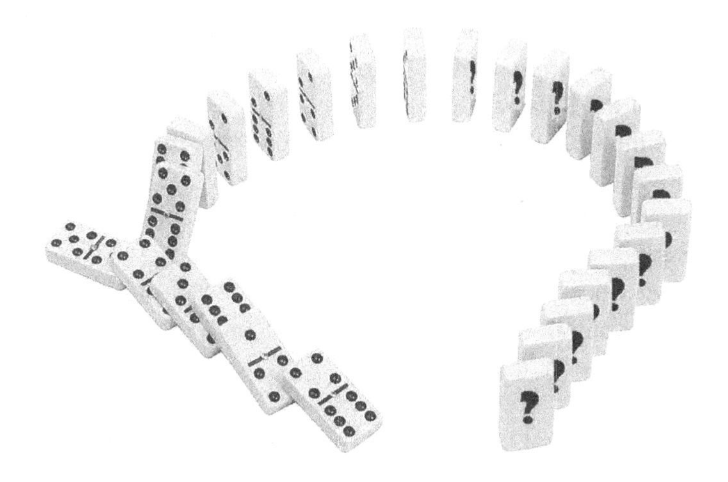

In this Appendix, we examine the issue of free will, discussed in Chapter 7, in more detail. For the purposes of this discussion, consider the game in which an arrangement of dominoes is set up and then, if the alignment is perfect enough, knocking over the first domino knocks down all the others.

Under strict determinism, if one domino hits a second, that second domino will fall. Thus, there is no 'free choice' on the part of the domino. If the lack of free choice means no free will, then dominoes have no free will. But suppose these are 'quantum dominoes,' and that the fate of each is genuinely indeterministic: each has a 50% chance of falling after being hit. According to the statistical laws of quantum theory, about half of the dominoes must fall, and the larger the total number of dominoes, the more precisely the fraction of fallen dominoes will approach ½. Some researchers have argued that this constraint effectively rules out free will as well, since if each domino could freely choose whether or not to fall, this seemingly would violate that statistical constraint. That is, why then couldn't all the dominoes fall, if they are really free to choose? On thc other hand, if each domino freely chooses such that the statistics happen to be obeyed every time, that would seem to be a miracle.

Thus, any account that presumes to retain free will based on quantum indeterminism must explain why that free will does not seem to violate the statistical constraints. There are answers to this challenge in the literature (e.g., Clarke, 2010), but here's one way to see how one can have a global quantum statistical constraint while still retaining free will at the individual level. I'll present the argument in terms of an allegory called *The Land of the Quantum Dominoes.*

The God of the Land of Quantum Dominoes decreed that all His creations, the Dominoes, would have free will. For direct interactions between the Dominoes themselves, the God made the outcomes of those actions fully indeterministic, such that the outcome of any such interaction had a 50% probability. By 'direct interactions,' for the purposes of this allegory, we mean that when one Domino fell on another, the second Domino had a 50% chance of falling down. Thus, the second Domino could freely choose whether or not to fall down.

However, as noted above, if a large number of Dominoes fell on their neighbors, about half of those neighboring Dominoes would have to fall in order to satisfy the Domino God's decreed laws. This implies that there was some unseen influence that affected all the Dominoes at some subtle level, such that when one neighboring Domino freely chose not to fall, he or she would, in effect, make it slightly harder for other neighboring Dominoes to choose to remain standing. Does this sound familiar? Recall that if we have two electrons in an atom, those electrons are subject to a collective constraint, the Pauli Exclusion Principle (PEP). If we think of one of the electrons as 'choosing' a particular position relative to the nucleus, its neighboring electron is influenced to choose a different position. This is a very real influence, the effect of which can be indirectly confirmed through measurement, even though there is no known physical mechanism for it.[1]

[1]In quantum theory, the effect is formally accounted for by noting that the electrons' quantum states are entangled. The entanglement causes the electrons to nonlocally influence each other through their quantum correlations, just as in the EPR experiment described in Chapter 2. A similar sort of influence affects entangled photons, but in the opposite direction: they tend to want to be in the same quantum state. Perhaps the latter better describes human collective behavior.

Thus, it is certainly logically possible that each individual Domino could have free will, but the extent of the freedom of each of its specific choices is constrained by the choices made by its fellow Dominoes. If the Dominoes' free will is a reflection of quantum uncertainty, their mental realm is characterized not only by that genuine indeterminism, but also by any quantum influences that go along with it. Just as electrons are subtly but unavoidably influenced by each other, the Quantum Dominoes are subtly but unavoidably influenced by each other as well.

Are we Quantum Dominoes? That is, are we subject to a kind of PEP that places unseen but influential constraints on our choices, even though each of those individual choices is fundamentally free? In this context, we might recall the psychological theory of Carl Jung, who proposed the idea of a 'collective unconscious.' The realm of the collective unconscious is analogous to what the Australian aboriginals called 'Dreamtime,' from which their myths were derived. If this notion of a 'collective unconscious' implies some sort of deep psychological connection among human beings, could this have any relevance to possible quantum correlations among humans? This book is primarily about physics, so we do not pursue that psychological topic here, but merely note it for those with the interest and expertise to pursue such a possibility, if they so (freely?) choose.

There's one more approach to consider here, and that is the option of simply abandoning free will. Such an approach has recently been advocated by some philosophers (e.g., Caruso, 2013). It is known as 'disillusionism,' meaning that we are supposed to recognize that our subjective sense of free choice is an illusion. Disillusionism is based either on a deterministic interpretation of quantum theory (such as the Bohmian interpretation) or on taking the quantum statistics as constraining our choices and actions so tightly that, in effect, those choices are predetermined. In that case, physically we are akin to dominoes that are being figuratively 'fallen on' by other dominoes. Each time that happens, whether or not we also will fall depends not on our own choices, which are not really free, but simply on the physical conditions of each fall. That is (figuratively speaking), sometimes one domino will fall on another, but the neighboring domino will not be knocked over, simply because the first domino was not quite close enough to the second one to overcome its inertia.

In this picture, all of our choices and actions are determined by circumstances and forces over which we have no control at all. Whenever we do anything, it is because we are compelled to do so. If one doesn't like the term 'compelled,' perhaps another word is 'propelled.' Whatever words we use to describe the situation, we are effectively automatons in which each fully-predictable input results in an equally fully-predictable and unavoidable output. This means that whenever we perceive ourselves as 'trying' to do something, it is in fact already decided whether our 'attempted' action will occur, and what its outcome will be. Therefore, in this disillusionist approach, isn't our subjective sense of 'trying' to do things also an illusion that would need to be rejected?

Returning to our deterministic dominoes: suppose those dominoes were sentient. While they might be able to perceive themselves as being involved in various processes and as exerting effort, in fact they are not self-propelled. Instead, they are propelled by forces beyond their control, since all their actions are fully dictated by those forces. So, in what sense are any of those dominoes really 'trying' to do anything? Every action that occurs is fully explained by physical processes and forces, so no 'trying' on the part of any of the dominoes is really part of the explanation for anything that occurs. If a domino perceives itself as exerting an effort, that must be just a byproduct of the actions in which he is determined to engage, and therefore just another aspect of the free will illusion. In this picture, 'trying' is superfluous, and any conscious entity is simply an automaton, even if perhaps a sentient one.

The point of the above is that we can't have it both ways: either (1) we have free will, in which case we can exert creative efforts through our own volitional capacity toward specific aims that we are trying to achieve, or (2) under disillusionism, we are simply automatons that don't actually try to do anything. We just fall, as dominoes, where we are propelled to fall, and our subjective perceptions that we are exerting creative efforts are just as illusory as our subjective sense that we have free will. However, 'disillusionism' is certainly not demanded by physical law. We can indeed be self-propelled, and although we certainly are subject to some forces beyond our control, it does not follow that we are primarily propelled by them. The effort we must exert to accomplish our chosen tasks could be just as real as our ability to make those choices.

Appendix C

The Pusey, Barrett, and Rudolph (PBR) Theorem

As mentioned in Chapter 9, Pusey, Barrett, and Rudolph proved that quantum theory cannot be considered a kind of 'shell game' in which there is a hidden property only approximately described by quantum states. In this appendix we will discuss that proof in more detail. In this discussion, we'll be making use of electron spin. Recall that an electron could be measured in a Stern–Gerlach (SG) apparatus with its magnet oriented in the vertical direction, so that the electron could be found to be spinning either 'up' (U) or 'down' (D). Suppose one of these electrons was found to be 'up,' and then allowed to continue on to another experiment. It would be described by the appropriate possibility triangle:

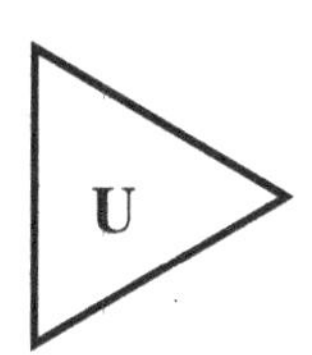

As before, let's represent this labeled triangle by the notation '| **U** >'. Now imagine that we have another SG box with its magnet oriented horizontally and to the right, so that an electron going through this device could end up spinning either 'right', | **R** > (which would be 'up' with respect to the horizontal direction), or 'left', | **L** > (which would be 'down' with respect to the horizontal direction). It so happens that the possibility | **U** > contains equal amounts of | **R** > and | **L** >.

Now suppose that we put our electron, prepared in state | **U** >, through that second measurement. If it is then detected at the left-hand detector, this means (in the usual way of understanding the theory, which neglects the absorption process) that its state has 'collapsed' to the possibility triangle labeled by 'L' (Figure C.1).

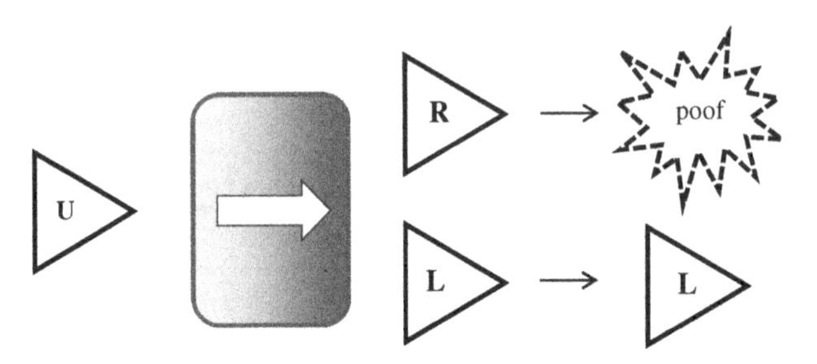

Figure C.1. The | **U** > state is measured and collapses to the state | **L** >.

Here is where we encounter the mysterious 'collapse of the wave function' that is inexplicable under the usual methods of approaching quantum theory: if the collapse is something that occurs in spacetime, it is clearly in conflict with relativity. That is because the result seems to propagate instantaneously; but nothing can propagate instantaneously in spacetime, because that is faster than the speed of light. This is the problem that is seemingly evaded by using the 'shell game' interpretation of the quantum state. Recall that the Shell Game Interpretation (SGI) states that the collapse was not something that happened to the quantum system; it was just our knowledge becoming more precise and focused, as if we pointed to one of the shells and found that the pea was not under it, thus reducing our ignorance about the location of the pea.

According to the SGI, that electron has some real physical property — let's use Einstein's term for this, an element of reality — that we can't readily get at. However, every time we measure it, the 'collapse' of the state just represents our knowledge becoming more precise, narrowing in on that hidden element of reality. That is why this type of interpretation is often called a 'knowledge interpretation.' Let's call the hidden property EOR for 'element of reality.'[1] The EOR is the 'pea' and the various possible quantum states are the 'shells.' The idea is that the electron really had property EOR all along (just as our handle was really located on the shelf in Aisle 12), but it could be labeled in less accurate terms by either by | **U** > or | **L** >, with some probability for each. That is, having the property

[1]Any resemblance of this term to a Winnie-the-Pooh character is unintentional, but interesting nevertheless.

EOR would predispose an electron to be found in | **U** > with some probability and to be found in | **L** > with some other probability. It's important to note that these probabilities are not part of standard quantum theory, which does not acknowledge anything like the EOR. They arise only in the SGI, in order to connect its proposed EOR to the usual quantum state. So to keep track of this point, let's call these EOR probabilities.

To set up the remainder of the PBR argument, imagine the hidden property EOR, proposed by SGI, as a 'masked man' that is hidden and inaccessible to measurement, but which secretly inhabits more than one possible quantum state, in this case both | **U** > and | **L** > (Figure C.2).

In the SGI approach illustrated above, the physically-real property of the system, EOR, is in the overlap between the two states | **U** > and | **L** > containing the 'masked man' (which represents EOR), while the empty areas of the states represent our ignorance of the actual physical situation. It's important to understand that EOR is neither | **U** > nor | **L** >; it is not a quantum state, just some alleged hidden property that is not directly detectable. In terms of our 'hardware store' analogy, the hidden property EOR is like the cabinet handle that we haven't found yet, and the quantum state descriptions such as | **U** > or | **L** > are like the different store areas ('hardware' or 'carpentry supplies') where it might actually be displayed. Using these basic concepts, the PBR theorem will show that there can be no such 'hidden' real feature of our quantum system; the 'collapse' cannot be interpreted as just our ignorance shrinking and our knowledge

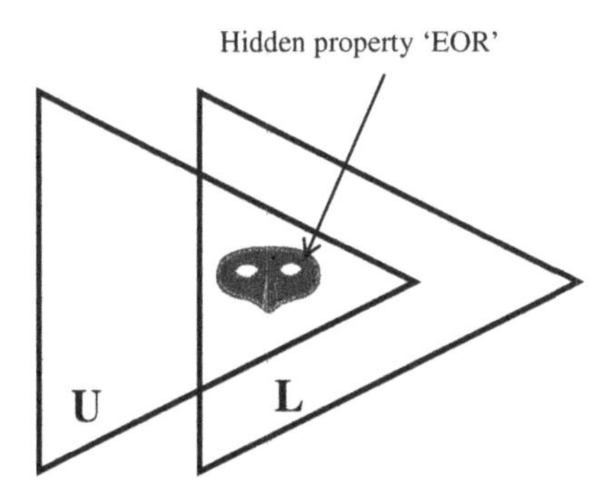

Figure C.2. The knowledge interpretation assumes there is some hidden element of reality, represented here by a 'masked man,' that is only approximately described by a quantum state such as | **U** > or | **L** >.

becoming more precise. In order to show this, we'll need to consider two electrons subject to the measurements discussed above, as well as some additional well-established features of quantum theory.

Remember that one SG device measured the 'vertical' aspect of the electron's spin (which could be either U or D), while the other one measured the 'horizontal' aspect (which could be either L or R). In order to do the proof, we need to use a system of two electrons, each of which could theoretically be prepared in either U or L. Let's represent these two-electron states by very long triangles, with patterns indicating the property U or L. The four two-electron possibilities are shown in Figure C.3).

Thus, the top left-hand triangle is the state for which both of the electrons are 'up'; the top right-hand triangle is the state for which electron 1 is 'up' and electron 2 is 'left'; the lower left-hand triangle is the state for which electron 1 is 'left' and electron 2 is 'up'; and the lower right-hand triangle is the state for which both electrons are 'left.' (You can also think of each of these tall triangles as two single-electron triangles for the relevant states 'stuck together,' but for our purposes below, it's better to represent the state of the two electrons as a single triangle representing both of the particles.)

Now, suppose that (unbeknownst to us) both electrons actually do possess the hidden state EOR, but we haven't yet done the SG measurements on them to find out whether each of them is 'U' or 'L.' (Each electron is measured by its own SG device, which can be set to measure either the vertical or horizontal direction.) According to the SGI, the various

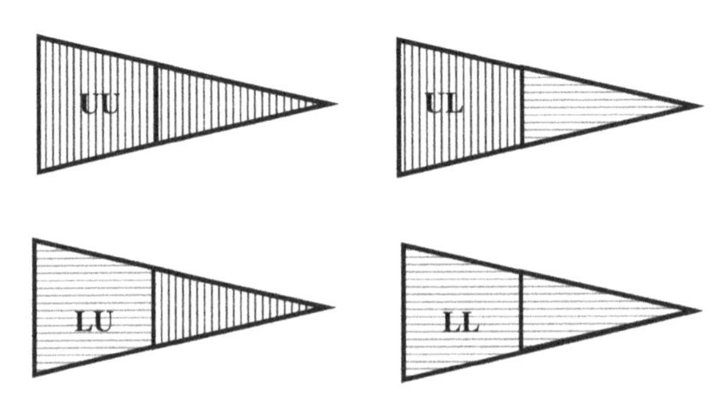

Figure C.3. The four two-electron states that can be composed of the one-electron states **| U >**, **| D >**, **| L >**, and **| R >**. Vertical stripes indicate the property U and horizontal stripes indicate the property L.

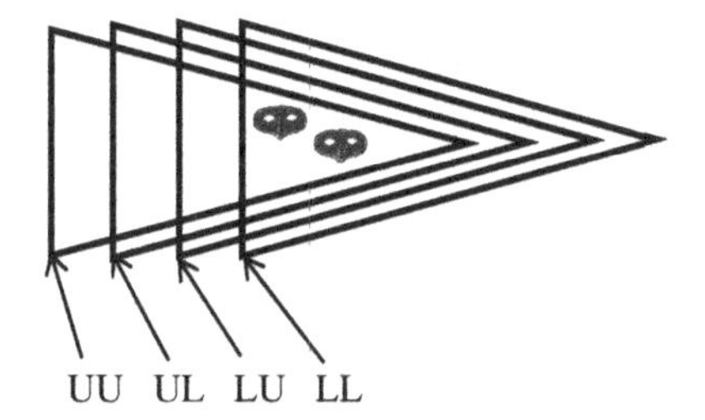

Figure C.4. Possible two-electron states for electrons both possessing hidden property 'EOR'. We omit the patterns in order to see the EORs clearly.

possibilities for the two-electron states must look like Figure C.4. That is, all the tall 'knowledge state' triangles must contain the hidden real states because, according to the SGI, there has to be some overlapping area — common to all these states — that contains the hidden properties actually possessed by each of the electrons. This is a crucial point, because it is what is meant by a 'knowledge interpretation': the idea that a quantum system really has some property and that the quantum state is just an approximate description that can be sharpened based on new information. This means that different approximate descriptions (UU, ... LL) can apply to the same hidden property, in the same way that there are two approximate, overlapping ways to describe the location of our cabinet handle in the hardware store (either by 'hardware' or by 'carpentry supplies'). According to SGI, these approximate triangle state descriptions are all we get when we do a quantum measurement; we cannot get at the exact element of reality, even if it is there.

For the next step in the proof, we need to consider an additional, different kind of measurement that we can make on the set of both electrons together. The states representing the possible outcomes of this next measurement are a bit more complicated to write down, but we don't need their explicit forms for the purposes of our discussion. The four possible outcomes of this particular measurement are basically the opposite properties of each of the above states. Let's therefore label those four possible outcome states as shown in Figure C.5.

It's important to note that whenever we do this second type of measurement, we must get one of these outcomes. And here's where quantum theory itself rains heavily on the SGI parade. As might be obvious from

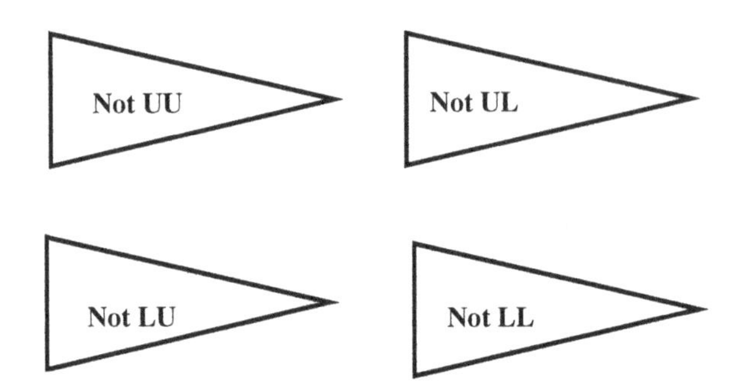

Figure C.5. The four possible outcomes of the second measurement that could be performed on both electrons. Whenever we do the measurement, we must get one of these.

the names of the states, the theory tells us the following when we perform this measurement on the two electrons:

> If the two electrons were in state | **UU** >, one would never get the outcome | ***NOT*** **UU** >.
> If the two electrons were in state | **UL** >, one would never get the outcome | ***NOT*** **UL** >.
> If the two electrons were in state | **LU** >, one would never get the outcome | ***NOT*** **LU** >.
> If the two electrons were in state | **LL** >, one would never get the outcome | ***NOT*** **LL** >.

This makes logical sense: if the two electrons are found to be in a certain state, regardless of what that state is, a measurement performed on them should never yield a result that directly contradicts that state. In fact, this is what we find in the ordinary SG measurement: if we input the state | **U** > into an SG device oriented in that same direction, we never get the opposite result, | **D** >.

Suppose both electrons happen to have some hidden state EOR, and they are first found in the quantum state | **UU** >, which means 'both electrons are up.' Visually, this would be represented by the left-most tall triangle in Figure C.4. Quantum mechanics (and basic logic) demands that if the two electrons were in this state and the second kind of measurement was then performed on them, they could not be found in the state | ***NOT*** **UU** >. This is because that state means precisely the opposite of

| **UU** >; i.e., that the electrons are not both 'up.' In order to be consistent with that, we have to say that when both electrons have the hidden property EOR they cannot have any chance of yielding the outcome | ***NOT*** **UU** > when this second kind of measurement is performed. This is because there is a possibility that when both electrons have the hidden property EOR, they might end up in the state | **UU** >. But if two electrons are in this state and then the second kind of measurement is performed, we must never be able to get the outcome | ***NOT*** **UU** >, which would contradict the state | **UU** >. So, in terms of what we called the 'EOR-probability' above, we must conclude that a set of two electrons both possessing the hidden property EOR must have zero EOR-probability of ending up the state | ***NOT*** **UU** > when the second kind of measurement is performed. This is because there is a chance that they might be found in the state | **UU** >, and if in that state, they can never be found in the state | ***NOT*** **UU** >.

Here's a way to see this in terms of the hardware store: again, the EOR is represented by the cabinet handle. In this store, there is a hardware section and a gardening section, and there are never any items in common between those two sections. Now suppose the handle could be stocked in home improvement. Then clearly it can never be found in the gardening section. So we must simply say that our handle has a zero EOR-probability of ever being found in the gardening section of the store.

But the same argument holds for all the other possible outcomes for the first kind of measurement (represented by the other three tall triangles in Figure C.4). For example, the hidden state could still be 'both electrons possess EOR' and the outcome found from doing the first kind of measurement could turn out to be | **UL** >. For consistency, the opposite outcome, '*NOT* UL,' cannot occur when the second kind of measurement is performed, so the EOR-probability for that outcome must also be zero. And so on, for all the possible outcomes of this second measurement. But this is absurd, because it leaves us with the following situation: given a real property EOR possessed by both electrons, and a legitimate measurement that could be performed on them (the second measurement we discussed), there is zero probability for any of the outcomes of that measurement to occur. This is nonsense, because we know that when we do the measurement, we will certainly get one of those outcomes.

Figure C.6. The shell game; the pea is the 'element of reality' and the shells are quantum states.

Here's a simpler way to visualize the gist of the proof: remember that in the shell game parable, the pea is the EOR and the white shells are the four possible quantum states that we get as outcomes when we perform the first kind of measurement (Figure C.6). But this magician can't cheat; he cannot remove the pea (just as, according to the key premise of the SGI, the quantum system really does have some hidden property EOR). Now suppose the magician swaps each of the original white shells with a black shell, in such a way that we still can't see where the pea ended up. It still has to be there, doesn't it? But, according to the proof, the pea has vanished. It cannot be under any of the black shells, or we will either contradict quantum theory or our own premises about quantum systems having a hidden property that could be found under any of the original white shells.

The problem here is caused by the idea that a real, but hidden, property EOR could result in more than one distinct quantum state, in this case both | **U** > and | **L** >. In terms of our shell game analogy, the problem is the idea that there is a pea that could be under any one of the shells (where the shells represent the quantum states). Remember that EOR is not a quantum state, but some property that supposedly could be common to two different quantum states, just like a cabinet handle could be classified as either 'hardware' or 'carpentry supplies.' This is what is meant by the idea that the 'collapse' of the quantum state describes not a real physical collapse but rather a sharpening of our knowledge about a quantum system. It implies that there is more than one possible state of knowledge about the same hidden truth, just as our knowledge of the cabinet handle's true location could be approximately described in different ways, or the pea might be found in any of the shells. The proof shows that quantum mechanics is not consistent with this common-sense idea, since it leads to a logical absurdity. We therefore must conclude that the quantum state does not describe our knowledge in the sense captured by the proof; it describes something with an indivisible uniqueness.[2] In terms of the shell game, there is no hidden pea. The 'shells' — the quantum states themselves — are the most precise descriptions of the true reality of quantum systems.

[2]Some proponents of 'shell game'-type interpretations argue that it is possible to retain the SGI by exploring loopholes in the proof. But this undermines the main motivation for the SGI as a common-sense way of avoiding collapse, since the loopholes involve stranger notions of what the underlying reality might be. One therefore faces a kind of 'diminishing returns' situation with this approach.

Bibliography

Abbott, E. (1884). *Flatland: A Romance of Many Dimensions.* London, Seeley & Co.

Anandan, J. (1997). 'Classical and Quantum Physical Geometry,' in Cohen, S. R., Horne, M., and Stachel, J. (eds.), *Potentiality, Entanglement and Passion-at-a-distance — Quantum Mechanical Studies for Abner Shimony*, Vol. 2. Kluwer, Dordrecht, pp. 31–52.

Bell, E. T. (1940). *The Development of Mathematics.* New York, McGraw-Hill.

Bohr, N. (1928). 'The Quantum Postulate and the Recent Development of Atomic Theory'. *Nature* 121, 580–590.

Bohr, N. (1961). *Atomic Theory and the Description of Nature.* Cambridge University Press, Cambridge.

Boltzmann, L. (1896). Vorlesungen über Gastheorie: Vol I, Leipzig, J.A. Barth; translated together with Volume II, by S.G. Brush, *Lectures on Gas Theory.* Berkeley, University of California Press, 1964.

Borges, J. L. (1941). *The Garden of Forking Paths.* Short story, trans. by Anthony Boucher (Aug. 1948) in *Ellery Queen's Mystery Magazine.* New York, Mercury Press.

Brown, H. (2002). *Physical Relativity.* Oxford, Oxford University Press.

Carroll, L. (1866). *Alice's Adventures in Wonderland.* New York, D. Appleton & Co.

Caruso, G. (2013). *Free Will and Consciousness: A Determinist Account of the Illusion of Free Will.* Lexington Books, Lanham, MD.

Clarke, R. (2010). 'Are We Free to Obey the Laws?' *American Philosophical Quarterly* 47, 389–401.

Cramer J. G. (1983). 'The Arrow of Electromagnetic Time and the Generalized Absorber Theory,' *Foundations of Physics* 13, 887–902.

Cramer, J. G. (1986). 'The Transactional Interpretation of Quantum Mechanics,' *Reviews of Modern Physics* 58, 647–688.

Davies, P. C. W. (1971). 'Extension of Wheeler-Feynman Quantum Theory to the Relativistic Domain I. Scattering Processes,' *Journal of Physics A: General Physics* 4, 836–845.

Davies, P. C. W. (1972). 'Extension of Wheeler-Feynman Quantum Theory to the Relativistic Domain II. Emission Processes,' *Journal of Physics A: General Physics* 5, 1025–1036.

Descartes, R. (1633). *Le Monde, ou Traite de la lumiere* (The World, or *Treatise on Light*). Trans. by Michael Sean Mahoney. New York, Abaris Books, 1979.

Descartes, R. (1641). *Meditations on First Philosophy*. Trans. by John Cottingham. Cambridge, Cambridge University Press, 1996.

Dyson, F. (2009). 'Birds and Frogs,' *Notices of the AMS* 56, 212–223.

Einstein, A. (1931) *Cosmic Religion: With Other Opinions and Aphorisms* 2009 edition. Mineola, NY, Dover Publications, p. 97

Eisberg, E. and Resnick, R. (1974). *Quantum Physics of Atoms, Molecules, Solids, Nuclei, and Particles*. New York, John Wiley & Sons.

Euler, L. (1770). *Elements of Algebra*. Royal Academy of Sciences, St. Petersburg.

Feynman, R. P. (1985). *QED: The Strange Theory of Light and Matter.* Princeton, NJ, Princeton University Press.

Flammarion, C. (1888). *L'atmosphère: météorologie populaire* Modern edition (French). France, Hachette Livre-BNF, 2012.

Gauss, C. F. 'Anzeige von "Theoria residuorum biquadraticorum, commentatio secunda",' *Gottingische gelehrt Anzeigen*, 23 April 1831.

Gribbin, J. (1984). *In Search of Schrödinger's Cat.* New York, Bantam.

Gribbin, J. (1995). *Schrödinger's Kittens and the Search for Reality*. New York, Little, Brown and Co.

Heisenberg, W. (1962). *Physics and Philosophy. The Revolution in Modern Science.* New York, HarperCollins.

Herbert, N. (1987). *Quantum Reality: Beyond the New Physics*. New York, Anchor.

Holevo, A. (1973). 'Bounds for the Quantity of Information Transmitted by a Quantum Communication Channel,' *Problems of Information Transmission* 9, 177–183.

Jammer, M. (1993). *Concepts of Space: the History of Theories of Space in Physics.* New York, Dover Books.

Kant, I. (1996). *Critique of Pure Reason*. English trans. by Werner Pluhar. Indianapolis, IN, Hackett,

Kastner, R. E. (2012). *The Transactional Interpretation of Quantum Mechanics: The Reality of Possibility.* Cambridge, Cambridge University Press.

Kastner, R. E. (2014a). 'Maudlin's Challenge Refuted: A Reply to Lewis,' *Studies in History and Philosophy of Modern Physics* 47, 15–20.

Kastner, R. E. (2014b). 'On Real and Virtual Photons in the Davies Theory of Time-Symmetric Quantum Electrodynamics,' *Electronic Journal of Theoretical Physics* 11, 75–86.

Kastner, R. E. (2014c). 'Einselection of Pointer Observables: the New H-Theorem?' *Studies in History and Philosophy of Modern Physics*. in press, DOI: 10.1016/j.shpsb.2014.06.004

Kuhn, T. (1962). *The Structure of Scientific Revolutions*. Chicago, University of Chicago Press.

Lucretius, 'On the Nature of Things.' Trans. by William Ellery Leonard (2008). (Online version, from Project Gutenberg.)

Marchildon, L. (2006). 'Causal Loops and Collapse in the Transactional Interpretation of Quantum Mechanics,' *Physics Essays* 19, 422–429.

Marlow, A. W. (ed.) (1978). *Mathematical Foundations of Quantum Theory*. New York, Academic Press.

Maudlin, T. (2002). *Quantum Nonlocality and Relativity: Metaphysical Intimations of Modern Physics Second Edition*. Hoboken, Wiley-Blackwell.

McTaggart, J. E. (1908). 'The Unreality of Time,' *Mind: A Quarterly Review of Psychology and Philosophy* 17, 456–473.

Newman, J. R. (1956). *The World of Mathematics*. New York, Simon and Schuster.

Norton, J. (2010). 'Time Really Passes,' *Humana.Mente: Journal of Philosophical Studies* 13, 23–34.

Petersen, A. (1963). 'The Philosophy of Niels Bohr,' *Bulletin of the Atomic Scientists* 19, 8–14

Polkinghorne, J. (1986). *The Quantum World*. Princeton, NJ, Princeton University Press.

Pusey, M., Barrett, J. and Rudolph, T. (2012). 'On the Reality of the Quantum State,' *Nature Phys.* 8, 475–478

Russell, B. (1959). *The Problems of Philosophy*. Oxford University Press, Oxford.

Schafer, L. (1997). *In Search of Divine Reality: Science as a Source of Inspiriation*. Fayetteville, University of Arkansas Press.

Stewart, I. and Golubitsky, M. (1992). *Fearful Symmetry: Is God A Geometer?* Oxford, Blackwell Publishing.

Tooley, M. (1997). *Time, Tense, and Causation*. Oxford University Press, Oxford.

Wheeler, J. A. and Feynman, R. P. (1945). 'Interaction with the Absorber as the Mechanism of Radiation,' *Reviews of Modern Physics* 17, 157–161.

Wheeler, J. A. and Feynman, R. P. (1949). 'Classical Electrodynamics in Terms of Direct Interparticle Action,' *Reviews of Modern Physics* 21, 425–433.

Wigner, E. (1960). 'The Unreasonable Effectiveness of Mathematics in the Natural Sciences,' *Communications in Pure and Applied Mathematics* 13, 1–14.

Index

Printed in the United States
By Bookmasters